Assessing Exposures and Reducing Risks to People from the Use of Pesticides

About the Cover

Strawberry harvesters during spring harvest near Santa Maria, California. The harvesters are protected from excessive exposure by safe field entry times, good personal hygiene, hand protection as a food safety and exposure reduction measure, and field toilets and sanitation facilities. (Photo coutesy of Helen Vega, University of California, Riverside).

ACS SYMPOSIUM SERIES **951**

Assessing Exposures and Reducing Risks to People from the Use of Pesticides

Robert I. Krieger, Editor
University of California at Riverside

Nancy Ragsdale, Editor
Agricultural Research Service
U.S. Department of Agriculture

James N. Seiber, Editor
Agricultural Research Service
U.S. Department of Agriculture

Sponsored by the
ACS Division of Agrochemicals

American Chemical Society, Washington, DC

Library of Congress Cataloging-in-Publication Data

Assessing exposures and reducing risks to people from the use of pesticides / Robert I. Krieger, Nancy Ragsdale, James N. Seiber ; sponsored by the ACS division of Agrochemicals.

     p. cm. — (ACS symposium series ; 951)

     "Developed from a symposium sponsored by the Division of Agrochemicals at the 229th National Meeting of the American Chemical Society, San Diego, California, March 13–17, 2005"—T.p. verso

     Includes bibliographical references and index.

     ISBN-13: 978–0–8412–3974–6 (alk. paper)

     ISBN-10: 0–8412–3974–6 (alk. paper)

     1. Pesticides—Toxicology. 2. Pesticides—Risk assessment. 3. Health risk assessment.

     I. Krieger, Robert Irving, 1943- II. Ragsdale, Nancy N., 1938– III. Seiber, James N., 1940- IV. American Chemical Society. Division of Agrochemicals. V. American Chemical Society. Meeting (229th : 2005 : San Diego, Calif.).

RA1270.P4A6   2006
363.738′498—dc22   2006042925

The paper used in this publication meets the minimum requirements of American National Standard for Information Sciences—Permanence of Paper for Printed Library Materials, ANSI Z39.48–1984.

Foreword

The ACS Symposium Series was first published in 1974 to provide a mechanism for publishing symposia quickly in book form. The purpose of the series is to publish timely, comprehensive books developed from ACS sponsored symposia based on current scientific research. Occasionally, books are developed from symposia sponsored by other organizations when the topic is of keen interest to the chemistry audience.

Before agreeing to publish a book, the proposed table of contents is reviewed for appropriate and comprehensive coverage and for interest to the audience. Some papers may be excluded to better focus the book; others may be added to provide comprehensiveness. When appropriate, overview or introductory chapters are added. Drafts of chapters are peer-reviewed prior to final acceptance or rejection, and manuscripts are prepared in camera-ready format.

As a rule, only original research papers and original review papers are included in the volumes. Verbatim reproductions of previously published papers are not accepted.

ACS Books Department

Contents

Indexes

Preface

Pesticides have an essential role in sustaining our bountiful food supply, protecting our property, and minimizing health impacts of disease vectors. Through the use of pesticides, modern agriculture has seen increased yields and more predictable food production, reduction in labor, and lower acreages in production to yield a given quantity of food. However, pesticides, by their very nature, are used to kill or interfere with the normal life cycles of living organisms that are classed as pests. This may give rise to ecological concerns when toxicity occurs in non-target organism, exposures exceed expected amounts, or drift damages offsite environments.

Many studies including estimates of human safety are required before a pesticide can be registered for use. The studies are designed to determine the parameters for using a given pesticide safely with minimum risk to human health or the environment. Because the dose makes the poison and the amount of exposure determines the dose, assessing human exposure to pesticides is critical in establishing the label requirements for use.

Effective risk management is based upon expert experience and scientific data including accurate determination of the determinants of pesticide exposure such as dose, sources, routes, and duration. Routes of exposure include dermal, inhalation, and ingestion. Direct exposures of pesticide handlers, harvesters of treated crops, and other agricultural workers are of first concern. Addition concern exists for bystanders inadvertently exposed to pesticides in the air from spray drift or from post-application volatization. In addition, long term, low-level exposures may result from ingestion of food and water. In determining overall risk, chronic and short-term (acute) exposures must be evaluated. Finally, the best available scientific data are marshaled to estimate potential exposures and possible health risks. Exposure reduction strategies are among risk mitigation measures. Other exposure reduction measures for pesticide handlers may include use of alternate products, formulations, application equipment and methods, personal protective equipment, special handling and disposal of remaining products and containers. Exposures of harvesters of treated crops are mitigated by clothing, hand protection (gloves and sound personal hygiene, and safe field entry (reentry) times.

This book is based on a symposium held at the National Meeting of the American Chemical Society (ACS) in San Diego, California, on March 14–15, 2005, honoring Dr. Robert Krieger, Cooperative Extension Specialist, University of California, Riverside, for his pioneering research in the assessment of chemical exposure and reduction of risks from the use of pesticides. Topics are grouped into exposure assessment, biomonitoring, and environmental measurements and mitigation. Chemical exposure assessment is examined from the perspective of risk assessment components as well as perceptions lacking scientific merit that influence policy. This book demonstrates that research and new refined assessment processes are paving the way for use of realistic estimates rather than worst-case default assumptions to establish regulations that result in reduced exposure risk.

The editors express their appreciation to the authors who submitted manuscripts that are included in this book, to the reviewers of those manuscripts (all manuscripts were reviewed by at least two anonymous peer reviewers), and to the ACS Division of Agrochemicals, which assisted in organizing the symposium program and funded some expenses associated with the program. Special gratitude is due the BASF Corporation for sponsoring the International Award for Research in Agrochemicals awarded to Dr. Krieger, which provided the occasion for the symposium upon which this book is based.

Nancy N. Ragsdale
George Washington Carver Center
U.S. Department of Agriculture
5601 Sunnyside Avenue
Beltsville, MD 20705–5139

James N. Seiber
Western Regional Research Center
Agricultural Research Service
U.S. Department of Agriculture
800 Buchanan Street
Albany, CA 94710

Chapter 1

Perceptions in Chemical Exposure Assessment

Robert I. Krieger

Personal Chemical Exposure Program, Department of Entomology,
University of California, Riverside, CA 92521

Chemicals used as pesticides are both a broad, vital catalyst for the support and advancement of all aspects of our lives, and at the same time targets of extensive suspicion and mistrust. Spectacular beneficial responses to chemical technologies in medicine, agriculture, nutrition, and manufacturing have occurred over long periods of time. Issues and common perceptions of the health and environmental significance of chemical exposure often dominate discussion of pesticide use indoors and in agriculture. As those technologies have been developed and used, adverse effects have been observed from time to time, but that reality is dwarfed by subjective feelings that often outweigh reason.

Chemicals

We live in a chemical world! The Chemical Abstracts Service now lists more than 22 million entries. The number increases every day. Only a small number of the 50,000 to 100,000 of them are chemicals of commerce, and pesticide active ingredients represent a much smaller number—perhaps 1,000 to 2,000. A still smaller number via inhalation, ingestion, and dermal contact are likely to contact humans and become part of our chemical experience. As more chemicals are added to the list, others are retired in a dynamic cycle. When all is said and done there can be little doubt that natural chemicals, principally in our diets, far outnumber our other chemical exposures.

Pesticides

No other group of chemicals known for their toxicity to pests is so extensively used as part of an attempt to maintain a balance of advantage over our competitors for food and fiber as well as vectors of disease. Pesticides as a group designate (a) any substance or mixture of substances intended for preventing, destroying, repelling, or mitigating any pest, and (b) any substance or mixture of substances intended for use as a plant growth regulator, defoliant, or desiccant. Pesticide use is associated with direct and indirect human exposure, however, when the amounts of exposure are well below those that might produce human effects it seems more appropriate to consider them chemical exposures rather than pesticide exposures since the exposure likely lacks either pesticide or toxicological activity (1). These exposures are measurable only with extremely sensitive analytical equipment, and they occur within an unnumbered and unmeasured chemical milieu. It was suggested that such detections be regarded as "trace ag(riculture) by-products, less than tolerance." At the present time, dose and time are often not distinguished in discussions of the occurrence and effects of chemicals used as pesticides, and as a result, unreasonable responses to chemical exposure can be expected, e.g. HAZMAT, drift, food.

No commercial use of chemicals is as controversial as pesticides (2). Persons in the United States are often strongly divided on grounds that are not easily defined (Table 1).

Table 1. Chemical Risk Characterization

Hazard Identification	Use
Dose-Response	Exposure Assessment
Risk Assessment	
Risk Mitigation	
Risk Communication	

A classification scheme derived from the risk assessment paradigm separates persons on whether their focus is "How much is too much?" or "How little is OK?" (Table 2). Hazard identification seems to be foremost among persons who forecast "all-or-none" responses with exposure, deny that dose is a determinant of response, and have little or no confidence in the scientific method as a means to predict human responses from animal studies. Other persons seem to integrate their feelings about pesticides with their general experiences with pharmaceuticals, food ingredients, beverages, and other chemicals of their daily lives. The Paracelsian truth that "Dose determines the

Table 2. Views of Chemical Exposures

How little is OK?	How much is too much?
Response	
"Dose makes the poison"	"All-or-none"
Amount	
Safe levels of everything	Small exposures cause certain harm
Laboratory Studies	
Awareness of limitations of toxicity testing in animals	Little confidence in relevance of toxicity testing

poison" establishes a foundation for their chemical encounters, and the idea that there is a safe level of everything is also consistent with our collective experience. Out of this perspective emerges a confidence in the scientific method as a means to meaningfully study responses of animals to protect human health.

It is unfortunate that the views held by persons who hold an "All-or-None" perspective are often very prominent in shaping public opinion about trace pesticide contaminants, especially when they occur in food and the public water supplies. These chemicals in tiny amounts could be considered to represent the chemical signature of the 21st Century but they are characterized as a "body burden." The following sections of this paper will overview some origins of the public's worrisome perception of pesticides.

Origins of Concern About Food Purity and Chemicals

Food Adulteration

The earliest public concerns about food purity were spawned by Fredrick Accum (3), a 19th Century chemist who addressed the adulteration of food. Pesticides were not an issue at this time, but food purity concerns were widespread and emotionally charged. Accum worked during a period of the emergence of many new chemical industries, and the appearance of quacks and impostors who made unlawful uses of new discoveries in chemistry. The adulteration of food and other necessities began to be practiced to an almost unlimited degree and in ways so subtle as to escape detection (4).

Accum was a teacher, tradesman, analyst, and technical chemist, but he is best remembered as an author of books about chemistry that appealed to the popular mind. The best known book of several written on foods was "Treatise on Adulteration of Foods." Accum discussed foods, their adulterants, and methods of detecting them in bread, beer, wine, spirituous liquors, tea, coffee, cream, confectionary, vinegar, mustard, pepper, cheese, olive oil, pickles, and other articles (4). In his crusade against food adulteration, Accum went beyond description of the frauds and indication of methods for their detection; he published the names of individuals who had been guilty of the practice. The cover of Accum's treatise carried the inscription "There Is Death In The Pot" and this philosophy became the foundation for the pure food movement based upon a quest for purity rather than findings shown to threaten human health.

Early Pesticide Residues

During the next 50 years (1850-1900) a larger national agriculture emerged, pesticide use became more common and concerns existed about possible health effects of fruit and vegetable pesticide residues. The chemicals of concern were primarily arsenicals.

A. J. Cook of Michigan reported results of the first official tests of arsenicals that considered consumer exposure in 1880 (5). Cook concluded that Paris green and London purple did not represent a danger to health. Eleven years later C. P. Gillette at the Iowa Agricultural Experiment Station also studied arsenicals on food and concluded that an individual would have to eat 30 cabbages dusted with Paris green to get enough arsenic to cause illness.

A more extensive residue survey was conducted 1915 to 1919 by the Bureau of Chemistry enforcing the Federal Foods and Drugs Act in response to intensified patterns of insecticide use. Hundreds of samples of peaches, cherries, plums, apples, pears, grapes, cranberries, tomatoes, celery, and cucumbers were tested for lead, arsenic, and copper. Little chemical residue remained on produce treated according to standard recommendations of the Department of Agriculture, but other samples treated with excessive amounts or too close to harvest had higher residues. The possibility of cumulative effects over a period of time also emerged in discussion of the significance of food residues at this time.

English, Canadian, and American orchardists faced trade and health concerns about the occurrence of lead, copper, and arsenic. The results of a British Ministry of Agriculture analysis of apple skin, stem, and calyx in 1925 are reported in Table 3. Results are reported as parts per million parts fruit. Seizures of contaminated pears occurred prompting litigation concerning whether or not the fruit "might be harmful to health (6)." The Royal Commission on Arsenical Poisoning offered 14 ppm (1/100 grain per pound) as all humans could tolerate. The Bureau of Chemistry adopted this "world tolerance" as a working standard for enforcement of the Food and Drugs Act.

Table 3. Lead, Copper, and Arsenic Trioxide Residues on Apples
circa **1925 and 2004**

Origin of Apples	Number of Samples	Lead (ppm)	Copper (ppm)	Arsenic Trioxide (ppm)
England[1]	13	0.06	0.2	trace
Canada[1]	6	0.9	0.2	0.4
USA[1]	5	0.4	0.3	0.2
Modern Total Diet Study[2]	-	0.002	0.2	0.003

[1]Recalculated from de Forest Lamb, 1936 (*6*)
[2]USDA, 2004 (*7*)

Western growers complained that they could not meet the 14 ppm tolerance, and the Secretary of Agriculture did not perceive an imminent threat to health. Leading toxicologists and physiological chemists were convened and after considerable study raised the tolerance to 35 ppm arsenic. They further reported

"--evidence as to the prevalence of lead and arsenic poisoning from the ingestion of fruits and vegetables sprayed with insecticides and fungicides is scanty and unconvincing, but inasmuch as the insidious character of accumulative poisoning by these substances causes such cases to be overlooked, the lack of evidence as to the prevalence of such poisoning must not be accepted as proof that instances do not exist."

These actions took place in an energized political climate with prevailing strong regional and international trade issues. Recognition of chronic lead poisoning in the industrial sector probably contributed to concern about accumulative poisoning from lead and arsenic residues on fruit and vegetable consumption. Current levels of these elements are reported (*7*) where arsenic and lead levels in market basket surveys were more than 2 orders of magnitude less than those reported by Lamb (*6*). Although time, sampling and analytical details are lacking, there can be little doubt that pest control practices at that time resulted in residues that would be considered unacceptable today.

20th Century Findings and Dr. Wiley's Poison Squad

Harvey Wiley (*8*) was a chemist and physician who served as Chief Chemist, U. S. Department of Agriculture. His famous Poison Squad of 12 employees who voluntarily lived in a boarding house where they were served meals containing what must have been maximum tolerated doses of food adulterants including boric acid, salicylic acid, sulphates, benzoates, and formaldehyde. Foods and excreta were collected, analyzed, and medical judgments were made about the impact of exposures on health. The work was of profound regulatory importance—Wiley is known as the Father of the Pure Food and Drug Act of 1906.

Pesticides were not included because they were present in small amounts and Wiley considered them a normal part of agricultural practice. However, one of the food adulterants studied by Wiley also has use as a residential insecticide. "It appears...that both boric acid and borax, when continuously administered in small doses for a long period of time...will create disturbances of appetite, digestion, and health." No further definition of "small doses" or "long periods" was made. It has been estimated that 500 mg/day was served up over 50 days to yield a daily dosage of 7 mg/kg-d. Wiley's work earned him the title "Old Borax." Recent concerns about toxicity of borate pesticides have resulted from developmental toxicity studies. NOAELs were each substantially greater in rats (<78 mg/kg), rabbits (125 mg/kg) and mice (248 mg/kg) than Wiley's dosage, including maternal toxicity in rats (163 mg/kg) and mice (250 mg/kg) (9). Occupational exposures are 0.07 to 0.3 mg/kg without adverse effects (10). Indoor structured activity on treated carpets gave exposures which were not distinguishable from exposures resulting from daily intake from fruits and vegetables ranging from 0.5 mg to 20 mg/day, and averaging 3 mg/day. The corresponding dosage was 0.02 mg/kg (11). The toxicological judgment remains no effect and the several sources of dose-response and exposure data seem to be complementary and supportive of current use.

Chemical Findings and Environmentalism

Some degree of persistence in a variety of environments is one of the important characteristics of chemicals that are suitable as pesticides. The measurement of those substances in a variety of environmental media including air, water, soil, and biota represents an important contribution to environmental studies made by trace chemical analysis. Unfortunately, for those persons who do not distinguish hazard identification and risk assessment, the triumphs of analytical chemistry have become the seeds for sustained concern and fear.

Several events that each had their roots in sensitive chemical analysis and concern about pesticide exposure seem to have been associated with public loss of confidence in chemicals for pest control. In chronological order the pesticide related issues included the 1959 Cranberry Scare, 1962 *Silent Spring* by Rachel Carson (12), 1989 Alar-Apple Fiasco, and continuing Workplace Chemical Exposures in pest management. Each of these situations illustrate continuing issues related to chemical exposures, pesticide use and public perception of unsafe chemical use in agriculture.

1959 Cranberry Scare

Thanksgiving dinner in 1959 in many households was celebrated without cranberry sauce. It became the first modern food scare supported by sensitive chemical analysis. The impact of the episode was compounded by inept risk communication. Earlier the government had announced that traces of the herbicide, aminotriazole, had been found in the cranberry crop of Washington and Oregon at 0.5-1% of the dietary level that caused thyroid cancer in rodents when fed for several years. At a press conference the Secretary of Health, Education & Welfare urged housewives "to be on the safe side" and refrain from buying cranberries because the rodent data suggested that the "contaminated" cranberries posed a human cancer risk. The comment stands as an example of incomplete risk communication. The comparable human dose was daily consumption of 15,000 pounds of tainted cranberries for several years; the risk of cancer if not zero, was very close to that.

The cranberry scare of 1959 signaled the modern wave of "chemical phobia" which persists to this day (Table 2). Many regulatory actions have been attributable to animal carcinogenicity testing using protocols that bear little relation to human exposure and cancer. The focus on cancer had its legal origins in the Delaney Clause, the 1958 amendment to the 1938 Food, Drug, and Cosmetic Act that prohibited the presence in food of any synthetic chemical that caused cancer in animal studies.

1962 Silent Spring

Public confidence in chemical technologies was further shaken by several events of the 1950s in addition to the cranberry episode. Public concern was heightened by reports of the occurrence of nuclear fallout from atmospheric testing, the thalidomide tragedy in Europe, the Torrey Canyon oil spill, and sanctions against chemicals as carcinogens can be listed. The impact of these events ultimately paled relative to the furor and immense influence of the publication of Rachel Carson's *Silent Spring*, a landmark book of the 20th Century. Carson stirred the world with her book charging dangers of pesticides. The book is regularly revisited by activists as though the original message was not understood. Activism must be recognized for focusing public attention, regulatory response and funding on some important and interesting aspects of modern environmental science and health. Unfortunately, it remains true that "good news" often isn't "news"!

Much misinformation is contained in Carson's book. She charged: "For the first time in the history of the world, every human being is now subjected to contact with dangerous chemicals, from the moment of conception until death."

8

The author continued: "500 new chemicals to which the bodies of men and animals are required to somehow adapt each year, chemicals totally outside the limits of biologic experience." Finally, Carson cast synthetic chemicals as "Elixirs of Death" related to the occurrence of DDT in humans and other living things. The exposures were perceived as contributing to a "load of toxic chemicals"—a biological dead end with uncertain consequences. The revival of the specter of a chemical "body burden" was perfectly suited to many of the chlorinated hydrocarbons and reflected earlier experience with lead in the workplace.

Silent Spring's most unlikely, but often quoted chapter, "A Fable for Tomorrow," presents a fictitious American town where all life -- from fish to birds to apple blossoms to human children -- had been "silenced" by DDT.

"There was once a town in the heart of America where all life seemed to live in harmony with its surroundings...a pastoral Eden of hardwood forests and bountiful wildlife...strange blight crept over the area and everything began to change...Everywhere was a shadow of death...It was a spring without voices. On the mornings that had once throbbed with the dawn chorus of robins, catbirds, doves, jays, wrens, and scores of other bird voices there was now no sound; only silence lay over the fields and woods and marsh...Even the streams were now lifeless...No witchcraft, no enemy action had silenced the rebirth of new life in this stricken world. The people had done it themselves..."

Action followed on all levels. President Kennedy appointed his scientific advisor to study the pesticide issue, and to produce a report containing recommendations for the use and regulation of pesticides in the United States. The President's Science Advisory Committee report, "The Use of Pesticides," issued on May 15, 1963, called for decreased use of toxic chemicals and use of chemicals that were less persistent in the environment. *Silent Spring* is generally credited or blamed with launching modern environmentalism. DDT became the prime target of the growing anti-chemical movement of the 1960s and 1970s that continues today.

It is easy to understand that the book infuriated many experts. Others were called to action. Subsequent attempts to develop a more informed public have failed on the grand scale of Carson's success. Most recently, *Silent Spring*, was acclaimed one of the top 100 contributions to journalism in the 20th Century by New York University's department of journalism.

Apples, Alar, and Another Challenge to Reasonableness

In 1989 the Natural Resources Defense Council (NRDC) and CBS-TV's *60 Minutes* manufactured a spurious national food scare by attacking the use and occurrence of Alar in apples. Anchorman Ed Bradley declared Alar "the single greatest cancer threat to children in the food supply." Bradley's conclusion has since been described as completely inaccurate by experts across the country. Others argued that the cancer risk was exceeded by eating a hamburger or a peanut butter sandwich. In the days that followed, the claims in the NRDC report were widely broadcast in the media. On his daytime talk show said Phil Donahue: "Don't look now, but we're poisoning our kids. I wouldn't lie to ya."

The media drive was unmistakable and soon a major frenzy existed over the wholesomeness of apples in the food supply. Michael Fumento (*13*), an insightful science and health author, covered Alar extensively in his book, *Science Under Siege*, in a chapter entitled, "The Alarm Over Alar." Fumento dismissed the charges against Alar for lack of evidence and further illustrated the power of the media. The following excerpts illustrate the power of the media and celebrity to shape public opinions about chemical technologies.

"... a consumer group decided to have a go at Alar. Apple growers liked Alar because it meant no premature dropping and even-sized fruit that could all be picked in one go. It passed its safety tests, and another set of tests when more stringent standards were introduced later.

Only one test proved equivocal – when a megadose of Alar was associated with cancers in mice, though not in rats. When the Natural Resources Defense Council activists got hold of this test, many years after it was published, they called it "new evidence." They promised the television show *60 Minutes* an exclusive.

Then the activist Ralph Nader telephoned the head of the department store Safeway and said "We're going to start a campaign to get Alar out of apples but why don't you save yourself a lot of trouble and us by saying that you're not going to buy any apples or apple products with Alar from your growers.

A week later Safeway put out a press release saying that they were buying no more products containing Alar. Then Nader telephoned the heads of other supermarket chains and told them that Safeway had stopped selling Alar-treated apples, and why not follow suit? They followed suit.

After the *60 Minutes* program, other journalists were given the press pack and ran their own stories. The public took the message to heart and, eager to prevent children from dying of leukemia, campaigned against Alar. The film actress Meryl Streep campaigned, as a mother, against Alar. So it was withdrawn.

The wholesale price of apples fell well below break-even level and put many growers out of business. When the new crops of Alar-free apples were ready, the retail price had rocketed. This was tough for mothers trying to feed children healthily on welfare handouts, though harmless to Meryl Streep and her children."

In this case NRDC, Consumers Union, the CBS-TV newsmagazine 60 Minutes and Ed Bradley, Ralph Nader, then-talk-show host Phil Donahue, and film star Meryl Streep captured the anxiety and heightened the fears of a wary public during the Alar incident of 1989. Public confidence in agriculture and the food supply was further diminished, and apple growers suffered major financial losses. Perceived pesticide hazards prevailed over dose-response data. An archaic cancer policy, once again, had impacts that were probably much greater than any of the assumed risks.

Chemical Foundations of Environmentalism

Food adulteration, pesticide residue analysis, aminotriazole residue on cranberries, DDT and Silent Spring, and Alar apples served to shake the confidence of consumers and promoted a perception of risk of unrealized health consequences. These events were important parts of the foundation for Environmentalism of the 1970s. Much remarkable set of legislation following on the heels of President Nixon's proclamation: "Clean air, clean water, open spaces—these should be the birthright for every American." "Freedom from risk" became a frequently heard expectation, and policy declaration of government. This position contrasts with the reality that safety is a matter of degree and not a concept subject to regulation by an *on-off* switch (Tables 1 and 2).

The most significant challenge facing pest management technologies may be to counter or outright replace the pesticide misperceptions that the public and some scientists and pesticide regulators have developed. . Litigation arising from miniscule exposure relative to harmful levels represents anxiety and fear carried by many persons.

Occupational Pesticide Exposures and Health

If we subscribe to the everyday realities of dose-response, there is a safe level of everything when dose, distribution, and use are considered. The public has accepted exposures to Botox, containing the most potent chemical known for blocking the release of acetylcholine; ethanol, sought in alcoholic beverages and common, but unrecognized, in produce and juices, and acetaminophen, the "go to" drug for parents and their babies in Tylenol, represent thousands of common chemical exposures. Our pesticide exposures are usually more dreaded, but often better studied (2). How are pesticide exposures to be treated within the dose-response paradigm?

The most useful point of reference for generalizations about the health significance of exposure is the work experience of pesticide handlers. Persons who mix/load/apply (handlers) have the opportunity for higher exposures than consumers. Exposures defined by (concentration χ time) easily distinguish handlers from consumers and bystanders (Table 4).

Factors that have safened the workplace during the past 70 years include generally lower hazard active ingredients, improved formulations, closed transfer systems, improved hose fittings and couplings, application techniques, personal protective equipment, and training to implement the Worker Protection Standard. Specific exposure mitigation factors cannot be assigned, due in part, to the pragmatic way that most advances have been developed and implemented, and the importance of personal hygiene can not be overestimated.

Physicians concerned with overexposure of organophosphate insecticide handlers and harvesters made the first critical assessments of worker exposure to modern organic pesticides. Griffiths et al. (14) monitored inhalation exposure of parathion applicators using respirator filter traps. Later Bachelor & Walker (15) reported dermal exposure after analysis of pads affixed to clothing during work. Subsequent detailed studies by Durham and Wolfe (16) established the "patch technique" that provided the large exposure database that is the foundation for the Pesticide Handlers Exposure Database (17) in North America. In recent times it is unfortunate that physicians are seldom a part of routine exposure assessment and risk characterization.

Table 4. Estimated Human Exposures Resulting From Selected Activities

Exposure Scenario	Organophosphate Dosage (ug/kg-d)	Source	Reference
Food	2.2-2.8	Potential dietary	Curl et al. 2003 (18)
	1.2	EPA dietary	Duggan et al. 2003 (19)
Workplace	6-14	Mix/load/apply	Krieger et al. 1998 (20)
	3-270	Malathion dust	Krieger & Dinoff 2000 (21)
Residential	0.0006-0.02	Drift and track-in	Krieger & Dinoff 2000
	1-30	Indoor foggers	Krieger et al. 2001 (22)
	0.2-1.3	Indoor broadcast	Krieger et al. 2001

Resident Exposures

Table 4 lists exposure estimates for unintended or unavoidable food, workplace, and residential exposures. The dosages range over more than 2 orders of magnitude and all are well within safe levels relative to a "toxic threshold" or the Lowest Observed Adverse Effect Level. For example, residential foggers produced exposures up to 30 ug/kg-d, but usually less than

10 ug/kg-d (22). Additionally, the residential "drift and track-in" biomonitoring (Table 4) occurred in a residence within the date gardens themselves where malathion dusters and harvesters worked (21). The exposures of the residents (that may or may not have come from the workplace) were 150 to 45,000-fold less than the no effect exposures of workers.

Risk Assessment

The chemical risk characterization paradigm offered more than 20 years ago by the National Research Council/National Academy of Sciences (*23*) is extensively used in the process of risk characterization or assessment (*24*) (Table 1). Hazard Identification is the determination of biological responses for the purpose of defining biological activities that may have relevance to human experience. The dose-response step defines the critical relationship between dose and fraction of a population responding to the stimulus.

The widespread adoption of the Risk Assessment paradigm (*23*) revealed a need for improved and refined pesticide exposure data. Whole body dosimetry using an inner garment to represent potential dermal exposure is such an intermediate advance (*25*). Biological monitoring when feasible, can improve the reliability of passive dosimetry, experimental determination of clothing penetration and dermal absorption, lowered detection limits, and longer monitoring periods are additional refinements. The range of human exposures is smaller than expected (Table 4) based upon subjective consideration of workplace activities. All are well below toxicity thresholds (lowest observed adverse effect level, LOAELs). It seems that the LOAEL is a more suitable guide than the No Observed Adverse Effect Level (NOAEL) divided by uncertainty factors (typically 100) to yield a reference dose (RfD). Scientific uncertainties that are not appreciated by some activists, media, some regulators, and the public result in misperceptions if RfDs are treated as disease thresholds.

The right side of Table 1 represents the Exposure element of the risk characterization process. Use is not usually represented as a separate category. Inclusion of "use" allows the risk assessor to clearly distinguish exposures by kinds of activities and may guide possible mitigation measures since the experience of mixer/loader/applicators and persons who reside in treated homes will usually be sharply different. The NOAEL/Exposure ratio yields the Margin-of-Exposure or a margin-of-safety. When MOE is factored by uncertainty factors representing individual variability (10x) and species-to-human variability (10x) the resulting reference dosage (RfD = MOE/100) can be estimated.

Subsequent steps in the risk characterization process include Risk Management and Risk Communication (Table 1). It is unfortunate that these processes imply an estimate of "risk" of the likelihood of exposure or without an estimate of the severity of illness. In fact, hazards only become risks when a susceptible population is exposed. Pesticide safety generally results from the conservative development of use patterns that minimize human exposure below experimental or epidemiological *no observed adverse effect levels factored by additional multiple uncertainty factors.* The resulting reference dose (RfD) causes some investigators, regulators, and members of the public to respond to exceedences of the RfD as though it represented a clinical end point that signals toxicity rather than as very conservative health guidance.

When safety evaluations were initially conducted, FIFRA was intended to "prevent unreasonable adverse effects on human health or the environment." Organophosphorous insecticides were introduced to California agriculture about 1950 accompanied by medical surveillance of cholinesterase analysis and urine biomonitoring overseen by physicians (Washburn, personal communication). With the passage of the Food Quality Protection Act of 1996 a still higher standard of safety prevails: "reasonable certainty of no harm," however, physicians seem to have a lesser role in the evaluation of the significance of pesticide exposure and the clarification of risk.

Aggregate exposure assessment using dietary food, water, and residential exposures have placed a premium on human exposure measurements that have become features of development, stewardship, and regulation of chemical technologies. The present system of study and ranking pesticide exposures and terming it "risk assessment" often fails to diminish public perceptions of hazard and may even heighten anxiety about normal pesticide exposures. Serious consideration should be given to increased participation of physicians and epidemiologists in the pesticide regulatory process to discern that toxicology *per se* is a small, but vital, part of assuring safe pesticide use.

References

1. Krieger, R. I.; Ross, J. H.; Thongsinthusak, T. Assessing Human Exposure to Pesticides. Rev. Environ. Contam. Toxicol. **1992**, *128,* 1-15.
2. Slovic, P. *Perceptions of Pesticides as Risks to Human Health*; Handbook of Pesticide Toxicology, ed. Krieger, R. I. Academic Press, San Diego, 2000; p. 845-857.
3. Accum, F. A Treatise On The Adulteration of Food, And Culinary Poisons. London: Longman, Hurst, Rees, Orme, and Browne, Paternoster Row. 1820.
4. Browne, C. A. The life and chemical services of Frederick Accum. J. Chem. Ed. **1925**, *2,* 829-851.
5. Porter, B. A.; Fahey J. E. Residues on Fruits and Vegetables in Insects The Yearbook of Agriculture, USDA, Washington, DC 297-301. 1952.
6. Lamb, R. de F. American Chamber of Horrors: The Truth About Food and Drugs, Farrar & Rinehart, New York. 1936.

14

7. *Total Diet Study Statistics on Element Results*. Revision 2, Market Baskets 1991-3 through 2003-4, USDA, Washington, DC. 2004.
8. Wiley, H. W. *The History of a Crime Against the Food Law*. Harvey W. Wiley, M. D., Publisher, 506 Mills Building, Washington, DC. 1911
9. Heindel, J. J.; Price, C. J.; Schwetz, B. A. The developmental toxicity of boric acid in mice, rats, and rabbits. Environ. Health Perspect. **1994**, *102*, 107-112.
10. Culver, B. D.; Smith, R. G.; Brotherton, R. J.; Strong, P. L.; Gray, T. M. 1994. *Boron*; Patty's Industrial Hygiene and Toxicology. Clayton, G. D.; Clayton, F. E. Eds. 4th ed. Vol. 2, Part F. Wiley-Interscience. John Wiley & Sons, New York, NY 1994 pp 4411-4448.
11. Krieger R. I.; Dinoff, T. M. Human disodium octaborate tetrahydrate exposure following carpet flea treatment is not associated with significant dermal absorption. J. Exposure Anal and Environ. Epidemiol. **1996**, *6*, 279-288.
12. Carson, R. *Silent Spring*, Boston: Houghton Mifflin, 1962.
13. Fumento, M. *Science Under Siege: How the Environmental Misinformation Campaign Is Affecting Our Lives*, William Morrow and Company, Inc., England. 1996.
14. Griffiths, J. T.; Stearns Jr., C. R.; Thompson, W. L. Parathion Hazards Encountered In Spraying Citrus In Florida. J. Econ. Entomol. **1951**, *44*, 160-164.
15. Bachelor, G. S.; Walker, K. C. Health Hazards Involved In The Use Of Parathion In Fruit Orchards Of North Central Washington. AMA Arch. Ind. Hyg. **1954**, *10*, 522-529.
16. Durham, W. F.; Wolfe, H. R. Measurement of the exposure of workers to pesticides. Bull. WHO **1962**, *26*, 75-91.
17. *Pesticide Handlers Exposure Database*, USEPA, Health and Welfare Canada, National Agricultural Chemicals Association, 1992.
18. Curl, C.; Fenske, R; Elgethun, K. Organophosphorous pesticide exposure of urban and suburban pre-school children with organic and conventional diets. Environ Health Perspect. **2003**, *111*, 377-382.
19. Duggan, A.; Charnley, G.; Chen, W; Chukwudebe, A; Hawk, R; Krieger, R. I.; Ross, J.; Yarborough, C.. Di-alkyl phosphate biomonitoring data; assessing cumulative exposure to organophosphate pesticides. Regulatory Toxicol. Pharmacol. **2003**, *37*, 382-395.
20. Krieger, R.I.; Dinoff, T. M.; Korpalski, S; Peterson, J. Protectiveness of Kleengard® LP andTyvek®-Saranex 23-P during mixing/loading and airblast application in treefruits. Bull. Environ. Contam. Toxicol. **1998**, *61*, 455-461.
21. Krieger, R. I.; Dinoff, T. M. Malathion deposition, metabolite clearance, and cholinesterase status of date dusters and harvesters in California. Arch. Environ. Contam. Toxicol. **2000**, *38*, 546-553.

22. Krieger, R. I.; Bernard, C. E.; Dinoff, T. M.;Ross, J. H.; Williams, R. L. Biomonitoring of persons exposed to insecticides used in residences. Ann. Occup. Hyg. **2001**, *45,* S143-S153.

23. *Risk Assessment in the Federal Government: Managing the Process;* National Research Council/National Research (NRC/NAS) Natl. Acad. Press, Washington, DC. 1983.

24. *An examination of EPA risk assessment principles and practices.* USEPA EPA/100/B-04/001, Risk Assessment Task Force, Washington, D.C. 2004.

25. Krieger, R. I.; Bernard, C. E.; Dinoff, T. M.; Fell, L.; Osimitz, T. G.;, Ross, J. H.; Thongsinthusak, T. Biomonitoring and whole body cotton dosimetry to estimate potential human dermal exposure to semivolatile chemicals. J. Exposure Anal. Environ. Epidemiol. **2000**, *10*, 50-57.

Chapter 2

Using Two-Day Food Consumption Survey Data for Longitudinal Dietary Exposure Analyses

Barbara J. Petersen[1], Stephen R. Petersen[2], Leila Barraj[1], and Jason Johnston[2]

[1]Food and Chemicals Practice, Exponent, Inc., 1730 Rhode Island Avenue, N.W., Suite 1100, Washington, DC 20036
[2]Durango Software, LLC, Bethesda, MD

Introduction

Consumer safety is a key component of pesticide registrations in all countries and in the establishment of maximum residue limits for pesticides. The procedures for estimating dietary exposures to pesticides have improved greatly in the past 20 years even as key methodology issues have emerged. In particular, the ability to reliably assess the impact of both acute and chronic exposures has been identified as a critical issue (1). The exposure assessment time frame should match the exposure period for the toxicological studies that were used to establish the reference dose. In practice for most chemicals, two different reference doses are considered: the ADI and the acute reference dose (aRFD). The ADI is established based on toxicology studies in which animals were dosed for chronic time periods while the aRFD is typically derived from studies in which exposures were short (at most a few days). The ADI should be compared to chronic exposure estimates and aRFD to acute exposure estimates.

Chronic exposure is particularly difficult to model since virtually all of the available surveys of food consumption are for at most a few days or a week. The most common approach to modeling chronic dietary exposure has been to estimate the mean daily consumption amount for each food/raw agricultural commodity (RAC) for all individuals in the population, multiply each RAC consumption amount by an average residue associated with that commodity, and sum the resulting products to obtain an average daily exposure, in mg/kg-day body weight. This approach does not estimate variability across consumers or across days for the same consumer. That is, this technique does not allow the estimation of exposures for those individuals who are often "heavy" consumers. Every individual in the population is assumed to have the same long-term exposure. In addition, there is no temporal basis for this exposure measure other than the ability to produce season-specific estimates of chronic exposure (i.e., separate estimates for each season of the year based on the time of year in which the consumption is estimated).

Considerations in Modeling Chronic Exposure

Computer-based exposure models have allowed major improvements in estimating acute exposure through the use of probabilistic assessment methods such as Monte Carlo analysis (6, 7). These models allow the analyst to make full use of the available data and provide estimates of the uncertainty in the estimate of exposures. A major problem has been the inability to simulate long-term exposures to residues in foods since none of the food consumption surveys capture quantitative data about dietary practices beyond a week or so. The National Health and Nutrition Examination Survey (NHANES) conducted by the US National Center for Health Statistics (NCHS) provides data that can be used to estimate longer term intakes for a small number of foods. Atlhough, there is little prospect of data on longer-term patterns becoming available for more foods (because of the extraordinary respondent burden, survey implementation difficulties and costs associated with such surveys), the data can be used to guide the selection of appropriate simulation models.

In this paper we evaluate alternatives for estimating chronic exposures. Quantitative information about food consumption patterns is essential for estimating consumer exposures. The use of the USDA's Continuing Survey of Food Intake by Individuals (USDA CSFII) for conducting acute (one-day) dietary exposure analyses has been widely accepted by risk assessors (6, 15 - 18). However, the consensus of these same risk assessors has been that two days of food consumption data for each individual are not a satisfactory basis for conducting longer-term (chronic and intermediate) exposure analyses (19). Similarly, Lambe and Kearney (20) warn against using short-term consumption data for estimating long-term or usual intakes and show that survey duration affect estimates of percent consumers, mean and high consumer intakes of foods, and the classification of individuals as high or low consumers of foods or nutrients.

In response to the FQPA, which mandated that the US Environmental Protection Agency (EPA) conduct cumulative and aggregate exposure analyses of pesticides from both dietary and residential sources, more complex "calendar-based" methods for calculating acute and chronic exposures have been developed which provide estimates of variability across individuals in a population and across days for the same individual. Several models[1] incorporating these methods are currently available. All of these models estimate exposures to contaminants in foods and facilitate the exposure assessment by estimating the intake of ingredients comprising the foods that were reported consumed by survey respondents in the USDA CSFII.

Two different methods have been developed for estimating variability within the population and for estimating intermediate- and long-term exposure analyses for individuals using the two-day food consumption records from the USDA CSFII: (1) the *"two-day repeated record"* approach uses the same two days of food consumption data repeatedly (but randomly drawn) for an individual for the duration of the exposure analysis, and (2) the "cohort record sharing" approach constructs cohorts of individuals with similar demographic characteristics (e.g., age, sex, ethnicity, region) and shares their food consumption records.

The *two-day repeated record method* has been considered simplistic and likely to "stretch the tails" of the distribution when used to estimate long-term exposure. For example, if an individual happens to eat one apple on each of the two days of the survey, then that individual will be assumed to eat one apple every day. Conversely, if that individual does not eat an apple on either of the two days in the survey, then that individual will be assumed to never eat apples. The *cohort record sharing approach* appears, at first sight, to provide a more representative sample of daily food consumption patterns for individuals with similar demographic backgrounds, resulting in a better estimate of the range of foods that individuals actually consume over time. However, this approach should result in individual food consumption patterns over longer time periods that are similar across individuals within a cohort, and so would not identify an individual who does indeed eat an apple every day. In fact, over a one-year period this approach could theoretically result in a mean exposure for a given population (approximately) equal to the "simplistic" chronic exposure that has been calculated in the past by simply multiplying the mean consumption amount for each commodity of interest by its average residue. Up until now, the determination of which approach provides a better simulation of typical dietary

---

[1] E.g., Calendex™
(http://www.exponent.com/practices/foodchemical/calendex.html), CARES®
(http://cares.ilsi.org/), and LifeLine™ (http://www.thelifelinegroup.org/)

practice has not been made. Therefore, research was undertaken to further evaluate these models.

The results presented below represent the completion of preliminary analyses presented at the ISEA 2004 meeting (22) and were obtained using both methods of extrapolating from two-day survey data (the two-day repeated record method and the cohort record sharing method). The results using each method are compared to results obtained directly in the NHANES food frequency survey. The comparisons were made for 51 of the 60 foods[2] that are included in the food frequency questionnaire in NHANES III for six U.S. subpopulations.

Data And Methods

Food Consumption and Food Frequency Data

The USDA CSFII (1994-96, 1998) was used to obtain information on 2-day food and nutrient intakes by 20,607 individuals of all ages, including intake data for 9,812 children from birth through 9 years of age. The USDA CSFII survey was statistically designed so that the results could be projected from the sample to the U.S. population. Survey participants were asked to provide socio-demographic and health-related information and information about their food intakes on 2 separate days. The 1988-94 NHANES III was used as the benchmark estimate of directly measured frequency of consumption of 60 types of foods. The list of foods is available at www.nhanes.gov along with estimates of the frequency distributions (data are also available at www.durango-software.com).

Models Evaluated

The Two-Day Repeated Record Method

The two-day repeated record method generates long-term consumption profiles for each "individual" in the USDA CSFII database by repeatedly

---

[2] A reliable way to compute frequency of consumption for some foods was not feasible (e.g., "any other vegetable," margarine, butter, and cooking oils) because of problems in definition, overlapping categories, etc. An analysis of the consumption of alcoholic beverages was not undertaken.

sampling from that individual's own two-day food consumption records. The approach has been implemented in Calendex™, a publicly available computer model for modeling dietary and aggregate exposure to pesticides and contaminants.

The Cohort Record Sharing Method

The cohort record sharing method matches "index" subjects to participants in the USDA CSFII based on a defined set of seasonal, socio-economic and demographic variables, and it builds the longitudinal consumption patterns by sharing person-days of consumption records from these matched subjects. The method has been implemented in Calendex™, CARES® and LifeLine™. The analyses reported here use the Calendex™ software. The cohort matching ability of Calendex™ is provided by a companion utility program called DMFgen (Dietary Matching File generator). The default matching criteria available in DMFgen include matching with respect to body mass index (based on three tertile groups), age (0-5 months, 6-12 months, 1-2 years, 3-5 years, 6-12 years, 13-19 years, 20-30 years, and 30+ years), and ethnicity (Hispanics, Non-Hispanic Whites, Non-Hispanic Blacks, and Others). DMFgen allows the user to define alternative demographic cohorts based on age and up to six other socio-economic, anthropometric and demographic factors (e.g., household income, region, sex, ethnicity, body mass index, and breast-feeding status for infants). Annual (365-day) dietary records for each of the 20,607 USDA CSFII participants with complete two-day intake records are created by randomly drawing daily consumption records from individuals in the matching cohort, preserving the day of the week (weekend versus weekday) and season. For the analyses presented in this paper, the default demographic specifications for DMFgen were used. Alternative demographic cohorts were defined using other criteria in order to test the sensitivity of the results shown in this paper to cohort definition, with no significant difference noted in the results.

Participants in the USDA CSFII report consumption of foods "as eaten" (e.g., pizza, apple pie, etc.). Calendex™ uses "recipes" prepared jointly by EPA and USDA to translate amounts of foods consumed from an "as eaten" basis to amounts of the corresponding RACs (i.e., tomato, apple, wheat, etc.). However, since the food frequency data available from NHANES III refer to groups of foods as eaten, the program was modified to allow it to use the original USDA data on foods "as eaten" so as to more closely align the NHANES & CSFII foods. Calendex was also modified so that it reported the frequency of consumption events for specific foods by each individual over the period of analysis, along with a distribution of these frequencies over the subpopulation of interest.

Results And Discussion

Thirty-day food frequency simulations were conducted for 51 different foods or food groups with both the two-day record repeat method and the cohort record sharing method (by season for latter method) using the modified Calendex program. Separate simulations were made for six different populations: males 12-19, females 12-19, males 20-49, females 20-49, males 50-99 and females 50-99. Actual 30-day food-frequency records for these same populations were calculated from the relevant NHANES data to serve as "benchmark" distributions. (NHANES food frequency records are not available for children under 12 so no comparison can be made for this age group. Note also that the NHANES frequency data are from a different survey and a different time period than the CSFII database used to simulate 30-day frequencies reported here, so we would not expect a perfect match in any case.)

Simulations for each of the 51 food groups were compared to the corresponding NHANES food-frequency data and tabulated for each of the six populations. The complete tabulation of results (51 food groups and 6 population subgroups) can be downloaded from the durango-software.com website. Table 1 presents these results for 15 different food groups, selected from the 51 food groups based on general interest, for males 12-19. For each food group six distributions are presented: the NHANES 30-day food consumption frequency (from the 10th percentile through the 99.9^{th} percentile), the corresponding two-day record repeat distribution, and four cohort record sharing distributions, one for each season. Figure 1 shows graphic results for four of the foods taken from Table 1.

The trend is very clear in the table and graphs: the two-day record repeat method tracks the NHANES frequency distributions much better at the top end of the distributions (generally above the 95th percentile and often above the 90th percentile) than does the cohort record sharing method, while the cohort record sharing method generally tracks the mid-range distributions (40-70%) with more accuracy. In most cases the cohort record sharing method provides estimates of intake frequency significantly lower than those made using the two-day record repeat method at all points above the 95th percentile.

Analysis of the entire data set (51 food groups x 6 subpopulations) shows that in only about 10% of the cases does the two-day repeat record method significantly overestimate the NHANES results at all points at or above the 99th percentile. Thus the use of the two-day record repeat method is unlikely to significantly overestimate food intake at the high end of the intake distribution in the majority of cases.

Based on these analyses, when the consumption patterns of high consumers (e.g. > 95th percentile) are desired, the *two-day repeat method* is more appropriate than the cohort record sharing method because the former better simulates the consumption patterns that are reported for frequent consumers by NHANES. On the other hand, the *cohort record sharing method* provides better estimates than the *2-day record repeating method* in estimating the central tendancies of the distributions of exposure.

The results reported here have focused on the frequency of consumption events. However, results of analyses conducted by Barraj et al. *(23)* in comparing the two-day repeat record method and the cohort record sharing method to actual seven-day intake levels of a selected number of foods (apples, bananas, breakfast cereals, fish, french fries, fresh tomatoes, and milk) derived from the UK Dietary and Nutritional Survey of British Adults *(5)* support the results of the current research. Specifically, the analysis showed that the two-day repeat method outperformed the cohort record sharing method for foods that showed a high level of within person correlation in amounts consumed (e.g., apples, breakfast cereals, and milk), while the opposite was true for foods that showed very little within person correlation in amounts consumed (e.g., fish and fresh tomatoes).

Conclusions

There is no entirely suitable substitute for a robust data base with daily food consumption diaries for thousands of individuals maintained over many weeks, months, and years. In fact, we have little more than two-day food intake records for about 20,000 individuals from the CSFII. But this short-term food consumption data can be used to simulate longer-term consumption trends for many commonly eaten foods. The two simulation methods examined in this paper are complimentary, and neither alone has the ability to predict the entire distribution of food consumption frequencies. At the high end of the frequency distribution (i.e., generally above the 95th percentile), the *two-day record repeat method* better simulates the directly measured consumption frequencies reported in the NHANES survey. The *cohort record sharing method*, performed poorly at the high end of the food consumption distribution. However, in the mid-range of the food consumption distribution (generally between the 40th and 70th percentiles) the cohort record sharing method provided more reliable estimates of food consumption frequencies relative to the two-day repeat method. Between the 70th and 95th percentiles, the actual frequencies of food consumption, and by extension the most realistic expectation for exposure values, are most likely bounded by the two estimates provided by these two simulation methods. Therefore, both of these approaches are useful. The risk assessor must select the most appropriate method for the intended purpose.

24

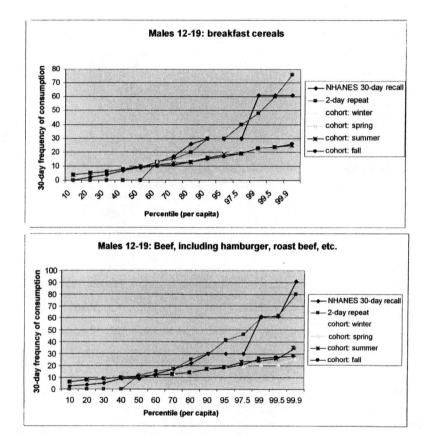

Figure 1. Graphical comparison of frequency of consumption distributions for four selected foods derived from NHANES III and simulated using the two-day repeat record method and the cohort record sharing method (Males 12-19 years)

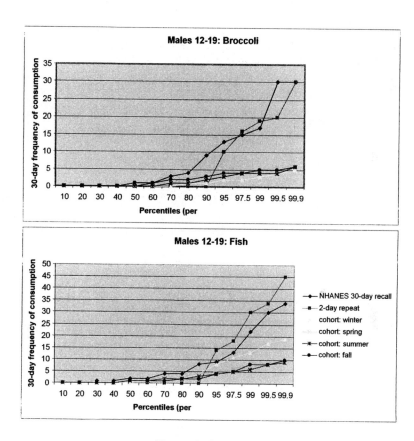

Figure 1. Continued.

Table 1. Tabular comparison of frequency of consumption distributions for 15 selected foods derived from NHANES III and simulated using the two-day repeat record method and the cohort record sharing method (Males 12-19 years)

Food	Age and Gender Groups	Percentiles													
		10	20	30	40	50	60	70	80	90	95	97.5	99	99.5	99.9
Breakfast Cereals: Except All-Bran, All-Bran Extra Fiber, 100% Bran, Fiber One, Total, Product 19, Most, and Just Right	NHANES III	0	2	4	7	9	13	17	26	30	30	30	61	61	61
	Repeat record method	0	0	0	0	0	13	15	20	30	30	40	48	60	76
	Record-sharing method-winter	4	6	8	10	11	13	14	15	17	19	21	23	25	26
	Record-sharing method-spring	5	6	8	9	10	10	11	13	15	17	19	20	23	26
	Record-sharing method-summer	4	5	6	8	10	11	12	13	16	18	19	23	24	25
	Record-sharing method-fall	4	5	6	8	9	10	11	13	15	17	19	23	24	26
Beans, lentils, and chickpeas/garbanzos, including kidney, pinto, refried, black, and baked beans	NHANES III	0	0	1	1	2	4	4	9	13	22	30	30	43	91
	Repeat record method	0	0	0	0	0	0	0	0	16	26	30	39	45	71
	Record-sharing method-winter	0	1	2	2	3	3	5	8	11	13	15	16	16	17
	Record-sharing method-spring	1	1	2	2	3	3	4	5	6	9	10	11	13	16
	Record-sharing method-summer	0	1	1	2	2	3	3	4	6	8	10	11	11	12
	Record-sharing method-fall	1	1	2	3	3	4	5	6	10	13	15	18	19	21

Food	Method														
Beef, including hamburger, steaks, roast beef, and meatloaf	NHANES III	3	0	0	0	9	13	17	22	30	30	30	61	61	91
	Repeat record method	0	0	0	0	12	15	17	25	30	41	46	60	62	80
	Record-sharing method-winter	5	8	10	10	11	12	13	15	17	18	20	21	21	26
	Record-sharing method-spring	8	9	10	11	12	13	14	15	17	19	21	23	24	27
	Record-sharing method-summer	6	8	9	10	11	12	13	14	17	19	23	24	26	35
	Record-sharing method-fall	6	8	9	10	11	12	13	14	17	18	21	26	27	28
Broccoli	NHANES III	0	0	0	0	0	1	3	4	9	13	15	17	30	30
	Repeat record method	0	0	0	0	0	0	0	0	0	10	16	19	20	30
	Record-sharing method-winter	0	0	0	0	0	0	0	1	1	2	2	3	4	5
	Record-sharing method-spring	0	0	0	0	1	1	1	2	2	3	4	5	6	6
	Record-sharing method-summer	0	0	0	0	0	0	1	1	2	3	4	4	4	6
	Record-sharing method-fall	0	0	0	0	1	1	2	2	3	4	4	5	5	6
Cheese, all types including American,	NHANES III	0	0	4	9	12	13	17	30	30	61	61	91	91	122
	Repeat record method	0	0	0	0	9	14	16	19	30	42	47	60	75	81
	Record-sharing method-winter	5	8	9	10	11	12	13	14	16	18	20	21	24	26

Continued on next page.

28

Table 1. Continued.

Food	Age and Gender Groups	10	20	30	40	50	60	70	80	90	95	97.5	99	99.5	99.9
Swiss, cheddar, and cottage cheese	Record-sharing method-spring	3	5	8	10	11	12	13	14	16	18	19	20	24	27
	Record-sharing method-summer	4	6	9	10	10	11	13	14	16	18	20	21	23	25
	Record-sharing method-fall	5	8	9	10	11	12	13	14	16	17	18	20	21	27
Chicken, all types, including baked, fried, chicken nuggets, and chicken salad. Including turkey.	NHANES III	1	2	3	4	4	5	9	9	13	17	26	30	30	61
	Repeat record method	0	0	0	0	0	0	14	16	28	30	39	49	60	62
	Record-sharing method-winter	3	4	5	6	8	9	11	13	17	20	21	23	24	27
	Record-sharing method-spring	3	4	4	5	6	6	8	10	14	17	20	23	24	28
	Record-sharing method-summer	3	4	5	6	6	9	10	11	13	14	16	17	18	19
	Record-sharing method-fall	3	5	6	6	8	9	10	12	14	16	19	21	23	25
Citrus fruits including oranges, grapefruits, and tangerines	NHANES III	0	0	0	1	3	4	9	12	17	30	30	61	61	91
	Repeat record method	0	0	0	0	0	0	0	0	0	12	19	30	32	64
	Record-sharing method-winter	0	0	0	1	1	2	2	3	4	6	9	13	17	19
	Record-sharing method-spring	0	0	0	0	0	1	1	2	3	4	5	6	8	9
	Record-sharing method-summer	0	0	0	0	0	0	1	1	2	2	3	4	4	5
	Record-sharing method-fall	0	0	0	1	1	1	2	2	3	4	4	5	6	9

NHANES III	0	0	1	2	2	4	4	8	9	13	22	30	34
Repeat record method	0	0	0	0	0	0	0	0	14	18	30	34	45
Record-sharing method-winter	0	0	0	1	2	2	2	4	8	11	13	17	19
Record-sharing method-spring	0	1	1	2	2	3	4	4	5	6	8	9	10
Record-sharing method-summer	0	0	0	1	2	2	3	3	4	5	6	8	9
Record-sharing method-fall	0	0	0	1	1	2	2	2	4	5	8	8	10

Fish including fillets, fish sticks, fish sandwiches, and tuna fish

NHANES III	2	4	4	8	9	13	17	30	30	61	61	122	182
Repeat record method	0	0	0	0	0	11	16	30	32	47	58	60	99
Record-sharing method-winter	3	4	5	6	8	9	12	15	21	28	31	35	36
Record-sharing method-spring	4	5	6	8	9	10	13	16	19	21	26	27	36
Record-sharing method-summer	4	6	8	9	10	11	13	16	18	20	23	25	28
Record-sharing method-fall	4	5	6	8	9	11	12	15	17	18	20	21	26

Other fruits such as apples, bananas, pears, berries, cherries, grapes, plums, and strawberries. Including plantains.

NHANES III	0	0	0	0	0	1	3	7	13	30	30	61	61
Repeat record method	0	0	0	0	0	0	0	0	0	16	19	30	30
Record-sharing method-winter	0	0	0	0	0	1	1	2	3	3	5	6	6

Peaches, nectarines, apricots, guava, mango, and papaya

Continued on next page.

Table 1. Continued.

Food	Age and Gender Groups	Percentiles													
		10	20	30	40	50	60	70	80	90	95	97.5	99	99.5	99.9
Peaches, nectarines, apricots, guava, mango, and papaya	Record-sharing method-spring	0	0	0	0	0	1	1	2	3	3	4	5	5	8
	Record-sharing method-summer	0	0	1	1	2	2	2	3	4	6	6	8	8	10
	Record-sharing method-fall	0	0	0	0	0	0	0	0	1	2	2	4	4	5
Rice	NHANES III	0	0	1	2	3	4	5	9	17	30	30	61	61	91
	Repeat record method	0	0	0	0	0	0	0	13	17	30	32	51	74	77
	Record-sharing method-winter	1	2	2	3	4	5	6	9	12	16	24	29	33	34
	Record-sharing method-spring	1	1	2	3	3	4	5	8	12	18	25	27	28	29
	Record-sharing method-summer	1	1	2	2	3	4	5	10	16	20	24	28	32	33
	Record-sharing method-fall	0	1	1	2	2	3	3	5	9	12	14	15	17	20
Spinach, greens, collards, and kale	NHANES III	0	0	0	0	0	1	1	4	7	13	17	30	30	30
	Repeat record method	0	0	0	0	0	0	0	0	0	0	12	16	18	45
	Record-sharing method-winter	0	0	0	0	0	0	1	1	2	2	3	4	4	6
	Record-sharing method-spring	0	0	0	0	0	0	0	0	1	2	3	4	4	4
	Record-sharing method-summer	0	0	0	0	0	0	0	1	1	3	4	4	5	5
	Record-sharing method-fall	0	0	0	0	0	0	1	1	2	4	5	8	8	8

Tomatoes including fresh and stewed tomatoes, tomato juice, and salsa	NHANES III	0	0	1	2	4	4	9	13	20	30	30	34	61	61
	Repeat record method	0	0	3	4	5	5	13	15	19	30	36	46	60	90
	Record-sharing method-winter	2	2	4	5	6	6	6	8	10	11	12	13	14	15
	Record-sharing method-spring	2	3	5	6	6	8	8	9	11	12	13	14	18	23
	Record-sharing method-summer	2	4	5	5	8	8	9	10	12	13	15	17	18	29
	Record-sharing method-fall	2	4	4	5	6	8	8	9	11	13	14	18	21	28
White bread, rolls, bagels, biscuits, English muffins, and crackers	NHANES III	4	9	13	17	22	30	30	30	61	91	91	122	152	243
	Repeat record method	0	13	16	28	30	34	44	48	60	72	77	90	99	108
	Record-sharing method-winter	15	23	25	28	30	31	34	35	39	42	45	47	47	50
	Record-sharing method-spring	18	23	25	27	29	31	34	36	43	43	45	48	51	62
	Record-sharing method-summer	20	25	28	29	31	33	35	36	40	42	43	45	46	49
	Record-sharing method-fall	24	26	29	31	32	34	36	39	41	44	46	47	50	53
White potatoes, including baked, mashed, boiled,	NHANES III	2	4	4	7	9	10	13	17	26	30	30	61	61	61
	Repeat record method	0	0	0	12	15	17	26	30	43	47	51	60	64	74
	Record-sharing method-winter	9	12	13	15	16	16	18	19	21	24	26	29	31	32

Continued on next page.

Table 1. *Continued.*

Food	Age and Gender Groups	Percentiles													
		10	20	30	40	50	60	70	80	90	95	97.5	99	99.5	99.9
french fries, and potato salad	Record-sharing method-spring	11	13	14	15	15	16	17	19	20	23	24	26	27	31
	Record-sharing method-summer	10	12	13	14	15	16	17	19	20	23	24	25	25	28
	Record-sharing method-fall	11	12	14	15	16	17	18	20	23	26	28	30	31	33
	NHANES III	0	0	0	0	0	0	1	2	5	9	13	26	30	61
	Repeat record method	0	0	0	0	0	0	0	0	0	0	14	18	30	48
Yogurt and frozen yogurt	Record-sharing method-winter	0	0	0	0	1	1	2	2	4	4	5	8	8	9
	Record-sharing method-spring	0	0	0	1	1	2	2	3	3	4	4	5	6	8
	Record-sharing method-summer	0	0	0	0	0	1	1	2	2	3	3	4	5	6
	Record-sharing method-fall	0	0	0	0	0	0	1	1	2	3	5	8	8	9

References

1. Hamilton, D. and Crossley, S. (2004). Pesticide Residues in Food and Drinking Water: Human Exposure and Risks, John Wiley & Sons Ltd, The Atrium, Southern Gate, Chichester, West Sussex PO19 8SQ, England, 2004.
2. WHO (1997). Guidelines for Predicting Dietary Intake of Pesticide Residues. ISBN. World Health Organization, Geneva, Switzerland
3. Pesticide Residues Committee. Pesticide Residue Monitoring Reports. Quarterly reports 2004. http://www.pesticides.gov.uk/uploadedfiles/Web_Assets/PRC/PRC_2004_Q1_text.pdf)
4. US Department of Agriculture (USDA) (2000). CSFII Data Set and Documentation: The 1994-96, 1998 Continuing Surveys of Food Intakes by Individuals. Food Surveys Research Group. Beltsville Human Nutrition Research Center. Agricultural Research Service. April 2000.
5. Gregory, J., Foster, K., Tyler, H. and Wiseman, M. (1990). The Dietary and Nutritional Survey of British Adults. HMSO, London, UK
6. US EPA (1998). Guidance for Submission of Probabilistic Human Health Exposure Assessments to the Office of Pesticide Programs (http://www.epa.gov/fedrgstr/EPA-PEST/1998/November/Day-05/6021.pdf, accessed 03/02/05)
7. Petersen, BJ, 2000. Probabilistic modeling: theory and practice. Food Addit. Contam 17(7):591-599.
8. Carriquiry, A. L., W.A. Fuller, J.J. Goyeneche & K.W. Dodd (1995). Estimation of usual intake distributions of ratios of dietary components. Dietary Assessment Research Series Report 5, Staff Report 95-SR 79. Center for Agricultural and Rural Development, Iowa State University, Ames, Iowa. 50011-1070.
9. Nationall Academy of Sciences. Subcommittee on Criteria for Dietary Evaluation (1986). Nutrient Adequacy: Assessment Using Food Consumption Surveys. National Academy Press, Washington, DC.
10. Nusser, S. M., A.L. Carriquiry, K.W. Dodd, and W.A. Fuller (1996). A semiparametric transformation approach to estimating usual daily intake distributions. J. Am. Stat. Assoc., 91, 1440-1449.
11. Slob, W. (1993). Modeling long-term exposure of the whole population to chemicals in food. Risk Analysis, 13, 525-530.
12. Slob, W. (1996). A comparison of two statistical approaches to estimate long-term exposure distributions from short-term measurements. Risk Analysis, 16, 195-200.
13. Institute of European Food Studies (IEFS) (1998). The Effect of Survey Duration on the Estimation of Food Chemical Intakes.
14. Tran, N., Barraj, l., Smith, K., Javier, A., and Burke T. (2004). Combining food frequency and survey data to quantify long-term

dietary exposure: a methyl-mercury case study. Risk Analysis, 24, 19-30.

15. US EPA (1999a). Acephate Revised Dietary Exposure Analysis (http://www.epa.gov/pesticides/op/acephate/rev_dietary.pdf, accessed 03/02/05)

16. US EPA (1999b). Chlorpyrifos Revised Acute Dietary Risk Assessment http://www.epa.gov/pesticides/op/chlorpyrifos/ acutedietary.pdf, accessed 03/02/05)

17. US EPA (2000). Available Information on Assessing Exposure from Pesticides in Foods. A User's Guide (http://www.epa.gov/fedrgstr/ EPA-PEST/2000/July/Day-12/6061.pdf, accessed 03/02/05)

18. 18. US EPA (2002a) Organophosphate Pesticides: Revised Cumulative Risk Assessment (http://www.epa.gov/pesticides/cumulative/rra-op/ I_C.pdf, accessed 03/02/05)

19. US EPA (2002b) Final report of the FIFRA Scientific Advisory Panel Meeting, February 5-7, 2002. A Set of Scientific Issues Being Considered by the Environmental Protection Agency Regarding: Methods Used to Conduct a Preliminary Cumulative Risk Assessment for organophosphate Pesticides (http://www.epa.gov/scipoly/sap/2002/ february/final.pdf, accessed 03/02/05)

20. Lambe, J., and Kearney, J. (1999). The influence of survey duration on estimates of food intakes – relevance for food-based dietary guidelines. British Journal of Nutrition, 81, Suppl. 2, S139–S142.

21. US Department of health and Human Services (DHHS) (1996). National Center for Health Statistics. Third National health and Nutrition Examination Survey, 1988 – 1994.

22. Johnston, J., Petersen, B., and Petersen S. (2004). A Data-Matching Algorithm to Generate Longitudinal Dietary Consumption Patterns. Presentation at the 2004 Annual Meeting of the International Society for Exposure Analysis.

23. Barraj, L., Walls, C., and Scrafford, C. (2002). Estimating Longitudinal Exposures Using Short-Term Data. Presentation at the Annual Meeting of the International Society for Risk Analysis.

Chapter 3

Percutaneous Penetration of Pesticides: Clinical Ramifications

Jackie M. Tripp[1], Francisca Kartono[2], and Howard I. Maibach[3,*]

[1]Division of Dermatology, Department of Medicine, University of British Columbia, Vancouver, Canada
[2]School of Osteopathic Medicine, Western University of Health Sciences, Pomona, CA 91766
[3]Department of Dermatology, University of California, 90 Medical Center Way, Surge 110, San Francisco, CA 94143

This chapter emphasizes the role percutaneous absorption plays in pesticide cutaneous and systemic toxicity. Discussion is limited to the use of pesticides in medicine, and the resulting toxic effects that have been observed. We focus on a brief introduction to clinical cutaneous adverse effects and penetration of pesticides – relating one to another. This chapter also summarizes assays designed to measure percutaneous absorption, which is a crucial step in assessing toxicity of a given chemical.

35

The *a priori* assumption that oral or inhalation exposures to pesticides are the most important is misleading. Homero et al. summarize minimal information on percutaneous penetration in Rhesus monkey. *(1)* The last generation's concept that the skin was not a major route of toxicity appears flawed. The cutaneous exposure route of pesticides can contribute significantly to total body burden and may have a greater impact on toxicity than ingestion or inhalation. Animal studies have shown that LD50 after cutaneous application can be on the same order as systemic dosing. *(2)* During spraying operations, skin deposition rates can reach higher values than obtained for respiratory exposure. *(2)* Diluents may alter pesticide penetration, Smith summarizes one part of this expanding knowledge - penetration enhancers. *(3)* The medical literature has significant information to offer with regards to pesticides, as they are commonly used in medicine to treat cutanaeous parasitic infections. One challenge in assessing the degree of toxicity of a particular substance is in devising the appropriate methodology for measuring and quantifying percutaneous absorption. Walker and Keith provides courtesy of the EPA, an efficient entry into human pesticide toxicity. *(4)* While animal models are readily available, in vivo techniques performed in humans are ideal and offer significantly more clinical relevance.

Use of Pesticides in Medicine

Pesticides are commonly used in medicine as pediculicides and scabicides; they include such ectoparasiticides as lindane, permethrin and malathion. Lindane (or hexachlorocyclohexane) is available as a shampoo or a lotion, permethrin (a synthetic pyrethroid) can be found in a 1% cream rinse or a 5% cream, and malathion (an organophosphate cholinesterase inhibitor) is commonly available as a 0.5% lotion. *(5)* While oral ingestion or inhalation may occur, because of the large body surface area (measuring 2 meters^2 in adults), the main cause of toxicity in the medical setting is percutaneous absorption. Potential reactions include local cutaneous reactions, as well as systemic toxicity. Highly toxic chemicals such as TCDD, azoxybenzenes, and dibenzofurans may be a passenger in commercially used pesticides. *(6)* The extensively studied chloracne stresses individual chemical potencies but no systematic studies on flux in man exist. *(6)* Agricultural workers and exterminators may have years of exposure, while medical uses are infrequent. Pesticide penetration in a physicochemical dependent manner is concentration dependent: some, like parathion, are linear from 4-4000µg/cm^2; others show a marked drop in linearity with increasing dose. *(6)* The greater the surface area, the greater the potential for increased flux. *(7)*

Local Effects of Pesticides

Reported local effects of pesticides include irritation, allergic contact dermatitis, photoirritation, photoallergic contact dermatitis, contact urticaria and subjective irritation. There is considerable overlap in the clinical appearance of these conditions, and their evaluation includes the use of a variety of provocative diagnostic tests available to the clinician.

Irritation

Irritation (irritant contact dermatitis) *(8)* is a complex multifaceted biologic process, with a diverse pathophysiology and clinical appearance. It appears clinically as red, itchy, sometimes painful areas on exposed skin. The exact mechanism of irritant dermatitis syndrome *(9)* is incompletely understood, but there is evidence that shows that its onset and development depends on such factors such as molecule characteristics, exposure time, and environmental conditions. Irritant contact dermatitis was initially thought to be primarily the result of a non-immunologic inflammatory reaction, but it seems that immunologic-like phenomena also occur. *(10)* Some pathophysiological changes that lead to irritation include skin barrier disruption, direct cellular epidermal damage, and the release of proinflammatory mediators. There is also increasing evidence that irritant chemicals exert some of their effect by interfering with the anti-oxidant system, increasing oxidative stress.

Pesticide irritant contact dermatitis may be acute, especially with spills of undiluted material, or cumulative. Penagos provides quantitative information on propensity of some agricultural chemicals to produce irritation. *(1)*

Allergic Contact Dermatitis

As with irritant dermatitis, allergic contact dermatitis (ACD) *(11)* is characterized by itching, redness, and other skin changes (edema, vesiculation). Whereas acute irritant dermatitis occurs on first contact with a cytotoxic chemical, ACD arises following more than one exposure to an allergenic chemical. The chemical penetrates the skin, sensitizes the immune system by generating T-lymphocytes that will respond to that chemical upon subsequent exposures. Upon that exposure, inflammatory mediators are released generating a clinical response.

The literature on pesticide allergic contact dermatitis is summarized by Penagos *(13)*; it is highly likely that most workers with this entity are not diagnosed because of lack of medical care. Reports on ACD have been noted on insecticides, soil fumigants, and fungicides along with other types of pesticides.

Photoirritation

Photoirritation (or phototoxicity) results from light interacting with a photoactive chemical resulting in a nonimmunologic irritation of the skin. *(12)* Prior to light exposure, the photoactive chemical can reach the skin either through topical application or through the circulatory system following ingestion. Its activation results in the release of inflammatory mediators such as histamine, kinins and arachadonic acid derivatives. The clinical response resembles an exaggerated sunburn and can feature redness, edema, vesiculation, desquamation, and pigmentary changes. We are unaware of currently utilized pesticides producing photoirritation in man. *(13)*

Photoallergic Contact Dermatitis

In general, photoallergic contact dermatitis (PACD) occurs less frequently than photoirritation. Light exposure of a photoactive chemical results in the formation of allergen, and sensitization to that allergen. On subsequent exposures an immunologically mediated inflammatory response is generated. Clinically this appears as an abrupt onset dermatitis of the light-exposed body regions, but this can spread beyond these areas and can even take on a chronic (or lichenoid) characteristic. Allegations of pesticide photoallergic contact dermatitis await clinical and photopatch test verification. (*13*)

Contact Urticaria

Contact urticaria (or immediate contact reactions) *(14)* appear within minutes after contact with the eliciting substance. The underlying pathophysiology involves mast cell release of inflammatory mediators such as histamine, prostoglandins, leukotrienes and kinins. This release can occur either as immunological contact urticaria (ICU) or non-immunological contact urticaria (NICU). The clinical reaction, which usually disappears within 24 hours, can range from itching and burning, to local wheal-and-flare reactions, to the uncommon phenomenon of generalized urticaria. In contact urticaria syndrome, a rarely encountered entitiy, there is such a strong hypersensitivity reaction that effects can be seen in other organs.

A dramatic example of agricultural chemical ICU was reported by Dannaker. (*15*) Minimal exposure in a sensitized nursery worker produced anaphylaxis.

Subjective (Sensory) Irritation

Sensory irritation refers to the symptoms of burn, sting and itch with exposure to some chemicals. Subjective or sensory irritation (*16*) can be seen after exposure to different chemical agents, but has been described specifically with pyrethroid insecticides. It occurs about 1 hour after contact, and peaks over 3-6 hours, and can last up to 24 hours. It is thought that these compounds exert this effect by penetrating the skin and interfering with axonal function. This symptom is sufficiently common with pyrethroids that premarket testing is widely performed. The human ear is especially sensitive and provides a model for its study with pyrethroids. Sinaiko et al. provides details. (*17*)

Diagnostic Tests for Cutaneous Effects

In vivo skin tests are used in dermatology to detect and define the possible exogenous chemical agent causing the skin disorder. The type of test is dependent on the type of reaction in need of evaluation.

Testing for irritation

Irritation cannot be accurately and consistently evaluated using any standardized in vivo diagnostic techniques. (*18*)

Testing for Allergic Contact Dermatitis

ACD is most commonly assessed with patch testing, as it is the most standardized of the testing methods. (*18*) Patch testing involves applying a patch over a panel of potential allergens on the patient's back, removing the patch 48 hours later and observing for any reactions, performing a second reading 24-48 hours after the removal of the patch, and occasionally observing the back one last time 1 week after the initial application of the patch. Success of the patch test is dependent on the experience and skill of the interpreter. The diagnostic patch test is a well-defined bioassay. Lachapelle (*19*) provides a concise and user-friendly how-to-do. It details patch size, limitations, difficulties, anatomic site, reproducibility and interpreter variability. Unfortunately there is no international commercial vendor – requiring the health care personnel to mix pesticides for this purpose.

ACD can also be evaluated with open testing and intradermal testing. Open testing involves looking for reactions at 7 and 28 days after repeated application of the potential allergen twice daily for 28 days. Compared with patch testing, open testing is not as sensitive. Intradermal testing can be helpful in some instances but involves intradermal injections, and in some instances carries a small potential risk of anaphylaxis, and hence is performed in centers with training in this assay.

Testing for Photosensitivity

For photosensitivity reactions photopatch testing should be performed. (*18*) Two patches of allergen panels are applied for 48 hours. After removal, one set is irradiated with ultraviolet light, and the other is protected. A reaction only at the irradiated site suggests photoallergy, reactions at both sites suggests a contact allergy, and reactions at both sites with a stronger reaction at the irradiated sites suggests a combination of both contact allergy and photoallergy.

Testing for Contact Urticaria

Evaluation options include the open, use, scratch and prick tests. (*14*) As mentioned above, appropriate training and facilities are required for patient safety.

Testing for Subjective (Sensory) Irritation

The lactic acid test has been used most commonly in the evaluation of subjective irritation. (*18*) The testing involves grading a patient's subjective sensations following exposure to sauna heat and lactic acid. The scoring system classifies patients as either "stingers" and "non-stingers", the former being more susceptible to subjective irritation.

Systemic Effects

Systemic effects have been elegantly documented by Krieger. *(6)* Lindane has been shown in case reports to be neurotoxic and hematotoxic resulting in seizures and aplastic anemia respectively. (*20*) This seems to be of concern mainly in the pediatric population and in patients with extensive skin disease. Due to these toxicity concerns, the use of lindane as a medical therapeutic has decreased. Malathion toxicity from percutaneous absorption is extremely rare, and systemic effects are seen with oral ingestion, where respiratory distress is the most concerning feature, similar to poisoning with other organophosphates. (*20*) Malathion is widely used on commercial crops and is rapidly inactivated. There have been no serious reported systemic adverse reactions to permethrin. (*20*) After decades of fatal percutaneous penetration after skin exposure, parathion use has been prohibited in California. (*1*)

Human exposure to these compounds has also been linked to sterility (*21*) and there have been extremely rare reports of fatal reactions. (*22, 23*) Sterility has so far only been observed in the occupational setting.

Interindividual Differences

Many factors influence the degree of penetration of a given substance through the skin. Some factors are more influential than others.

Anatomic Site

Obvious differences exist in skin absorption resistance as a function of anatomic site. (24) Most data on penetration is discussed in terms of forearm penetration, but penetration through forearm skin does not always reliably predict absorption at all anatomical sites, and may underestimate toxicity resulting from exposure to other areas. Each anatomical site has its own calculated penetration index (P_i), with the forearm being the reference point and the scrotum being the most barrier deficient. For pesticides, the arm had a penetration index of 1, with the other body surface areas having P_i values as follows: 1 for the leg, 3 for the trunk, 4 for the head, and 12 for the genitals. Forearm Total Body Exposure (FTBE) estimates whole body penetration by extrapolating data from forearm penetration, but a more accurate approach would be to use the Potential Total Body Exposure (TBE), which takes into account the different P_i values of different body areas. For pesticide exposure, the calculated TBE is more than double that of the FTBE, since the different P_i values from the different areas are factored into the calculation.

Gender

There does not seem to be any significant differences in intrinsic skin characteristics between the genders. (25) Gender-based differences in the frequency of contact dermatitis is most likely due to gender-influenced differences in exposure. Although fat stores are in general greater in women than men, dermatopharmaceutical studies in man have not been done in enough women to document gender difference with pesticides. (26)

Ethnicity

Transepidermal water loss often mirrors penetration of the body; thus groups (including ethnic groups) can have estimates of increased or decreased flux based on the readily performed water loss measurements. (26) When compared to white skin, black skin has higher transepidermal water loss, lower skin surface pH, and larger mast cell granules. Experiments on the properties of Asian and Hispanic skin demonstrate contradictory results, and do not seem to show any significant differences. Other properties such as skin water content, corneocytes desquamation, skin elastic recovery/extensibility, lipid content and skin microflora show some statistically significant differences between races, but overall these results are minimal and contradictory. In sum, racial or ethnic background does not likely play a role in percutaneous absorption. (27)

Age

Both irritant and allergic inflammatory reactions are weaker in older patients. (28) It is not clear why this occurs, but it could be due to an age-related decrease in percutaneous absorption or an age-related difference in the inflammatory cascade.

Newborns, on the other hand, especially preterm neonates, have immature epidermal barriers, which can lead to potential problems with percutaneous absorption of toxins. (29)

Methods for Measuring Interindividual Differences in Percutaneous Absorption

Experimental models can be used to help determine absorption, and specifically demonstrate interindividual differences.

Correlating Transepidermal Water Loss and Percutaneous Absorption

Transepidermal water loss (TEWL) is the outward diffusion of water through the skin and it is used to evaluate skin water barrier function. (*30*) The TEWL is obtained by using an evaporimeter that measures the pressure gradient between the skin surface and ambient air. TEWL is an indicator of flux across the skin barrier, and most studies investigating TEWL and percutaneous absorption, seem to indicate that the two are indeed correlated, although the precise qualitative nature of this correlation has yet to be elucidated. This relationship between TEWL and percutaneous absorption allows for the TEWL measurements to be used as a predictive model in animals and humans.

Corneocyte Surface Area

The laws describing diffusion through a membrane assign a certain importance to membrane thickness. However, when examining the skin, one sees regional variation in permeability that does not seem to related to stratum corneum thickness. It has been postulated that regional variations in corneocytes surface area play a role in percutaneous absorption. (*30*) With a larger corneocyte surface area, there is a smaller intercellular volume. The intercellular space acts as a molecular "reservoir", and the smaller it is, the less absorption of a given molecule. Corneocyte surface area appears correlated to flux and hence offers a surrogate marker for measuring individual differences in pesticide penetration. *(30)*

Stratum Corneum Mass

Adhesive tape studies have been used to calculate percutaneous absorption profiles by inducing barrier disruption prior to permeability assessment. Studies involve repeatedly tape stripping the skin and analyzing the tapes via colorimetric protein assay to determine the amount of stratum corneum removed. Additionally, the TEWL is measured after a certain interval of tape stripping to help correlate loss of stratum corneum thickness with functional properties of the barrier. There is evidence to suggest that there is marked interindividual differences in barrier disruption after tape stripping as measured by the TEWL. This variation may be a result of individual differences in response to the tape stripping injury. (*31*)

Methods for Measuring Percutaneous Absorption

Methods exist for determining the percutaneous absorption of topically applied chemical compounds.

Measuring Radioactivity of Excreta or Blood

Measuring radioactivity of excreta or blood following topical application of a labeled compound is the most common method of determining percutaneous absorption in vivo. This method doesn't account for metabolism by the skin, and can be quite labor intensive, but is nonetheless a reliable approach to measuring absorption. (*32*)

Advanced Analytic Technology

Recent advances in physical methods of analysis have allowed for detecting elements at increasingly minute amounts, often reaching below the parts-per-billion (ppb) level. (*33, 34*) By using these advances researchers can analyze skin absorption of chemicals without resorting to radionucleotides.

Inductively Coupled Plasma-Mass Spectrometry (ICP-MS)

ICP-MS measures characteristic emission spectra of ions produced by a radio-frequency inductively coupled plasma using optical spectrometry. The sample compound is first nebulized, and then transported to the plasma torch where it is ionized, and then analyzed. The absorption of a given chemical can be determined quantitatively by analyzing the amount of that chemical in body fluids using this method.

Stable-isotope ICP-MS analysis

A potential confounding factor in assessing metal absorption data with ICP-MS is inadvertently measuring not only elements absorbed through the skin, but also picking up naturally occurring trace elements, as well as elements in the diet. Through the use of ICP-MS to analyze in vivo artificially generated stable metal isotopes, it is now possible to differentiate between the amount of chemical absorbed through the skin, and that absorbed with the diet. Ultimately, this allows for measurement of skin absorption of chemicals, independent from endogenous natural isotopes present in the body.

Inductively Coupled Plasma-Atomic Emmission Spectroscopy (ICP-AES)

ICP-AES is an analytical method that detects metals in trace amounts. It is less sensitive than ICP-MS, and can be used to detect levels above 1 ppm. At these levels, ICP-MS would be overwhelmed, leading to instrumental problems.

Particle Induced X-Ray Emission (PIXE)

PIXE employs a proton beam which activates an atomic electron elevating it to a higher orbit. When an outer shell electron falls back to fill the vacancy, the transition is measured as the emission of an X-ray photon. This technique exhibits sensitivity approaching 0.1ppm, and can be used to analyze ultra-thin strips of stratum corneum removed by tape stripping.

Simplified In Vivo Penetration Assay

Studies have shown that the total mass of a chemical within strippings of the stratum corneum after 30 minutes of application time is directly correlated to the urinary flux of the compound at 4 days. (*35*) In other words, the simple measurement of a chemical within the stratum corneum at the end of 30 minutes of application gives a good predictive assessment of the total amount penetrating the skin within 4 days.

In Vivo Direct Assays in Humans

Measuring breath samples is a non-invasive means to study dermal absorption. Exhaled breath can be analyzed using an ion-trap mass spectrometer (MS/MS) equipped an atmospheric sampling glow discharge ionization source (ASGDI). (*36*) The intensity data collected by the mass is converted to concentration values in ppb. This method has great potential as a non-invasive real time method of determining bioavailability of organic solvents following dermal exposure, without significant lag time.

Taken together, these advanced tools greatly improve our ability to quantify penetration with almost all agricultural chemicals.

Conclusions

In summary, considerable evidence provides documentation for the cutaneous and systemic toxicity from cutaneous pesticide exposure. Systemic and topical toxicity (dermatotoxicity) relate to both degree of flux and potency. Thus malathion and parathion have vastly different systemic effects in man – with parathion producing death. This relates more to potency than flux. *(6)* Although methods for ranking toxicity potency exist, much of the data is either not published, or not generally available. Hopefully, next generation cooperation between physicians, government and industry will lead to information that will effectively prevent cutaneous and systemic effects.

References

1. *Pesticide Dermatoses*, Penagos, H.; O'Malley, M.; Maibach, H.I.; CRC Press: Boca Raton, 2000.
2. Maibach, H.I.; Feldman, R.J. 1974. Systemic absorption of pesticides through the skin of man. In *Occupational Exposure to Pesticides: Report to federal working group on pest management from Task Group on Occupational Exposure to Pesticides,* Appendix B, pp. 120-127. US Government Printing Office, 1975, 0-551-026, Washington D. C.
3. *Percutaneous Penetration of Enhancers*, Smith, E.; Maibach, H.I.; Taylor & Frances: Boca Raton , 2005.
4. *EPA's pesticide factsheet data base,* Walker, M.M.; Keith, L.H.; Lewis Publishers: Chelsea, 1992.
5. Roberston, D. B.; Maibach, H. I. In *Basic and Clinical Pharmacology*, 9th ed.; Katzung, B. G., Ed; McGraw-Hill: New York, 2004; pp 1021-1022.
6. Wester, R.C.; Maibach, H.I. In *Handbook of Pesticide Toxicology*, 2nd ed.; Robert I. Krieger ed., Academic Press, San Diego, CA, Vol. 1, Chap. 43, pp. 905-912, 2001)
7. *Topical Absorption of Dermatological Products,* Bronaugh, R.L.; Maibach, H.I. Marcel Dekker Inc.: New York, 2002.
8. Weltfriend S.; Ramon M.; Maibach, H.I. In *Dermatotoxicology*, 6th ed.; Zhai, H.; Maibach, H.I., Eds.; CRC Press: New York, 2004; pp 181-228.
9. *Irritant Dermatitis Syndrome*, Chew, A.; Maibach, H. I., Eds.; Springer: New York, 2005.
10. Dika, E.; Branco, N.; Maibach, H. I. Immunologic patterns in allergic and irritant contact dermatitis: similarities. *Exog. Dermatol.* **2004,** *3*, 113-120.
11. Marzulli, F. N.; Maibach, H. I. In *Dermatotoxicology*, 6th ed.; Zhai, H.; Maibach, H.I., Eds.; CRC Press: New York, 2004; pp 229-234.

12. Marzulli, F. N.; Maibach, H. I. In *Dermatotoxicology*, 6th ed.; Zhai, H.; Maibach, H. I., Eds.; CRC Press: New York, 2004; pp 341-352.

13. Penagos, H. A. In *Pesticide Dermatoses*, Penagos, H.; O'Malley, M.; Maibach, H. I., Eds.; CRC Press: NewYork, 2001; p 23-39.

14. Amin, S.; Lahti, A.; Maibach, H. I. In *Dermatotoxicology*, 6th ed.; Zhai, H.; Maibach, H. I., Eds.; CRC Press: New York, 2004; pp 817-848.

15. Dannaker, C. J.; Maibach, H. I.; O'Malley, M. Contact urticaria and anaphylaxis to the fungicide chlorothalonil. *Cutis* **1993**, *52*, 312-315.

16. Herbst, R. A.; Strimling, R. B.; Maibach, H. I. In *Toxicology of the Skin*; Maibach, H.I., Ed; Sheridan Books: Ann Arbor, MI, 2000;pp105-113.

17. Sinaiko, R.; Maibaich H.I. Bioengineering Correlates of the Sensitive Skin Syndrome: The Sensory Irritation Component. In *Bioengineering of the Skin. Water and the Stratum Corneum*, 2nd Ed., Fluhr, J.; Elsner, P.; Berardesca, E.; Maibach, H.I.; eds, CRC Press: Boca Raton, Chapter 16, pp 187 - 196, 2004

18. Amin, S.; Lauerma, A.; Maibach, H. I. In *Dermatotoxicology*, 6th ed.; Zhai, H.; Maibach, H. I., Eds.; CRC Press: New York, 2004; pp 1007-1020.

19. *Patch Testing and Prick Testing. A practical guide*, Lachapelle, J.M.; Maibach, H.I., Springer-Verlag: Berlin, 2003.

20. Rasmussen, J. E. In *Comprehensive Dermatologic Drug Therapy*; Wolverton, S.E., Ed.; W.B. Saunders: Philadelphia, PA, 2001; pp 537-546.

21. Swan, S. H.; Kruse, R. L.; Liu, F.; Barr, D. B.; Drobnis, E. Z., Redmon, J. B. Semen quality in relation to biomarkers of pesticide exposure. *Environ. Health Perspect.* **2003**, *111*: 1478-1484.

22. Walker, G. J.; Johnstone, P. W. Interventions for treating scabies. *Cochrane Database Syst. Rev.* **2000**, CD000320.

23. Pannell, M.; Gilbert, J. D.; Gardiner, J.; Byard, R. W. Death due to malathion poisoning. *J. Clin. Forensic Med.* **2001**, *8*, 156-159.

24. Guy, R. H.; Maibach, H. I. In *Topical Absorption of Dematological Products*; Bronaugh, R. L.; Maibach, H. I., Eds.; Marcel Dekker: New York, 2002; pp 311-315.

25. Modjtahedi, B. S.; Modjtahedi, S. P., Maibach, H. I. The sex of the individual as a factor in allergic contact dermatitis. *Contact Dermatitis* **2004**, *50*, 53-59.

26. Levin, J.; Maibach, J.I. The correlation between transepidermal water loss and percutaneous absorption: an overview. *Journal of Controlled Release*, **2005**, 103: 291-299.

27. Wesley, N. O.; Maibach, H. I. Racial (ethnic) differences in skin properties: the objective data. *Am. J. Clin. Dermatol.* **2003**, *4*, 843-860.

28. Harvell, J. D.; Maibach, H. I. Percutaneous absorption and inflammation in aged skin: a review. *J. Am. Acad. Dermatol.* **1994**, *34*, 1015-1021.

29. Kravchenko, I.; Maibach, H. I. In *Neonatal Skin: Structure and Function,* 2nd ed.; Hoath, S. B.; Maibach, H. I., Eds.; Marcel Dekker: New York, 2003; pp 285-298.

30. Rougier A.; Lotte, C.; Maibach, H. I. In *Percutaneous Absorption,* 3rd ed.; Bronaugh, R. L.; Maibach, H. I., Eds.; Drugs and the Pharmaceutical Sciences; Marcel Dekker: New York, 1999; Vol. 97, pp 117-132.

31. Bashir, S. J.; Chew, A. L.; Anigbogu, A.; Dreher, F.; Maibach, H. I. Physical and physiological effects of stratum corneum tape stripping. *Skin Res. Technol.* **2001,** *7,* 40-48.

32. Wester, R. C.; Maibach, H. I. In *Percutaneous Absorption,* 3rd ed.; Bronaugh, R. L.; Maibach, H. I., Eds.; Drugs and the Pharmaceutical Sciences; Marcel Dekker: New York, 1999; Vol. 97, pp 215-227.

33. Hostynek, J. J.; Maibach, H. I. Advanced methods measure skin penetration at the parts-per-billion level. *Cosmetics Toiletries Magazine* **2005,** *120,* 30-33.

34. Hostynek, J. J.; Maibach, H. I. Advanced methods measure skin penetration at the parts-per-billion level: part II. *Cosmetics Toiletries Magazine* **2005,** *120,* 40-45.

35. Rougier A.; Dupuis, D.; Lotte, C.; Maibach, H. I. In *Percutaneous Absorption,* 3rd ed.; Bronaugh, R. L.; Maibach, H. I., Eds.; Drugs and the Pharmaceutical Sciences; Marcel Dekker: New York, 1999; Vol. 97, pp 375-394.

36. Poet, T. S.; Thrall, K. D.; Corley, R. A.; Hui, X.; Edwards, J. A., Weitz, K. K. et al. Utility of real time breath analysis and physiologically based pharmacokinetic modeling to determine the percutaneous absorption of methyl chloroform in rats and humans. *Toxicol. Sci.* **2000,** *54,* 42-51.

Chapter 4

Testing for Persistent Organic Pollutants in Banked Maternal Serum Specimens

William M. Draper[1], Jennifer Liang[2], Mary Fowler[3],
Martin Kharrazi[4], F. Peter Flessel[3], and S. Kusum Perera[1]

[1]California Department of Health Services, Sanitation and Radiation
Laboratory Branch, Richmond, CA 94804
[2]Impact Assessment, Inc., 2166 Avenida De La Playa, Suite F,
La Jolla, CA 92037
[3]California Department of Health Services, Environmental Health
Laboratory Branch, Richmond, CA 94804
[4]California Department of Health Services, Genetic Diseases Branch,
Richmond, CA 94804

Human blood sera from the California Maternal Serum
Expanded Alpha-Fetoprotein (XAFP) prenatal screening
program were analyzed for persistent organic pollutants
(POPs) in a pilot biomonitoring study. POPs, including DDT
compounds (DDTs), chlorobiphenyls (CBs) and brominated
diphenyl ethers (BDEs), were determined in the less than two
mL specimens using a dual capillary column gas
chromatography-electron capture detector (GC-ECD) method.
Twenty-six target compounds were determined including 6
DDTs, 15 CBs and 5 BDEs, BDE-47, -99, -100, -153, and
-154. Among 40 XAFP specimens accessioned between May and
June, 2002 in three southern California counties, 4,4'-
DDE was detected in all with a range of 0.17 to 8.9 ng/mL.
4,4'-DDT was detected in only two subjects. 2,2',4,4'-
Tetrabromodiphenyl ether (BDE-47) was found in 55% of
serum specimens, but only two sera had elevated sBDE (the
sum of 5 BDEs determined) with 1.4 and 1.9 ng s-BDE/mL.
This study demonstrates that chemical analysis of
environmental contaminants in banked XAFP specimens is
technically feasible, and that chemical analysis of these
specimens could provide exposure information in population-
based studies.

One of the fundamental hurdles in biomonitoring is obtaining specimens in a cost effective and safe manner that protects the privacy and other interests of human subjects. Collecting blood, for example, requires access to people in a clinical setting. Drawing blood is invasive and uncomfortable, and must be done by a licensed phlebotomist. These impediments and associated costs have limited the ability of public health researchers to employ biomonitoring as a means to identify exposed groups, or even establish normal reference ranges to important environmental contaminants.

The objective of this investigation was to examine the potential for chemical analysis of exposure markers in banked maternal serum specimens collected in the 15th to 19th week of pregnancy. Sera were obtained from the California Maternal Serum Expanded Alpha-Fetoprotein (XAFP) program, a state-run prenatal screening program involving private regional laboratories. Participation rates are very high, i.e., 70 to 80% of pregnant women in California elect to participate in screening. Thus, XAFP specimens represent a large, randomly drawn cross section of the population. These specimens also provide information on vulnerable segments of the population including women of child bearing age, the developing fetus and breastfed children. An earlier study of environmental tobacco smoke (ETS) exposure was based on cotinine levels in XAFP specimens (1).

While XAFP sera and other human specimens have scientific value in public health research and epidemiologic studies (2, 3), it is not known whether these specimens are appropriate for trace analysis of environmental pollutants. XAFP samples are collected in polypropylene serum separator tubes (SSTs) that contain a blood clot activator and gel additives. Artifacts such as phthalates from plastic containers or uncontrolled sample handling could render the specimens useless. Another limitation of XAFP specimens is their small volume, only one to two mL of serum remain after prenatal screening.

The present study focuses on three important classes of environmental chemicals, chlorinated pesticides, chlorobiphenyls (CBs) and polybrominated diphenyl ether flame retardants (BDEs). All of these compounds are persistent organic pollutants (POPs) known for a long biological half-life and association with adipose tissue and blood lipids. There is considerable literature on chlorinated pollutants in human serum, and they continue to be of interest even though some of them were banned for domestic use over three decades ago. DDT exposure continues to be important among California's many immigrants from Mexico, Central and South America and Southeast Asia where DDT is still used (4). Brominated flame retardants (BFRs) are emerging environmental contaminants only recently detected in human blood and adipose and wildlife specimens from California (5).

In this study a multiple residue, dual capillary column gas chromatography-electron capture detector (GC-ECD) method was specifically developed for analysis of POPs in human serum. The validity of the method was established by analysis of a certified standard reference material. Finally, banked XAFP specimens drawn in three California counties in 2002 were analyzed as a test of feasibility.

Methods and Materials

Chemicals

Pesticide standards were obtained from the U.S. Environmental Protection Agency (USEPA) Repository for Toxic and Hazardous Materials, Environmental Monitoring and Support Laboratory (Cincinnati, OH) as neat standards or 1,000 to 5,000 μg/mL solutions in methanol. The standards were >99% pure and were not corrected for purity. BDE standards were obtained from Cambridge Isotope Laboratories (Andover, MA, URL http://www.isotope.com) as 50 μg/mL solutions in nonane. CBs were obtained as mixtures (CLB- 1) from the National Research Council of Canada, Institute for Marine Biosciences (Halifax, Nova Scotia, Canada, URL http://www.imb.nrc.ca). Four mixtures are provided, CLB-1 A through D, but only mixture D, containing the most abundant environmental CB congeners, was used. Individual CB congeners were obtained from Foxboro Company (North Haven, CT).

Solvents including methanol, hexane (95% n-hexane) and 2,2,4-trimethylpentane (isooctane) were organic residue analysis grade and were obtained from Mallinckrodt-Baker (Phillipsburg, NJ) as was granular, anhydrous sodium sulfate (ACS grade, 12-60 mesh). Diethyl ether (99+% , ACS reagent grade) and anhydrous benzene (99.8%) were obtained from Aldrich Chemical (Milwaukee, WI). Silica gel (230 mesh) was obtained from Sigma Chemical (St. Louis, MO). Laboratory reagent water was produced by a Barnstead/Thermolyne Nanopure Infinity UV water purifier (Dubuque, IA) that polished distilled feedwater with ion exchange, charcoal, and UV light.

Reference Human Serum and Calf Serum

A certified standard reference material (SRM) of human serum, SRM 1589a, was obtained from the National Institute for Standards and Technology (Gaithersburg, MD, URL http://www.nist.gov/srm). SRM 1589a is intended for use in evaluating analytical methods for the determination of selected CB congeners and other chlorinated pollutants, and is provided as a freeze-dried powder that is reconstituted in distilled or HPLC-grade water. The material has both certified and reference concentration values for 25 CBs, 10 chlorinated pesticides and other persistent and bioaccumulating compounds. Method accuracy and precision were evaluated by repeated analysis of SRM 1589a reconstituted in laboratory reagent water. Blanks, laboratory reagent water substituted for serum, also were analyzed repeatedly to evaluate laboratory contamination and bias. Calf serum used for method development was obtained from Invitrogen (Carlsbad, CA).

Serum Separator Tubes and Cryovials

Serum separator tubes (SSTs) were obtained from BD Vacutainer (Franklin Lakes, NJ). Cryovials were purchased from Corning Inc. (Acton, MA).

Instruments

The gas chromatograph used was an Agilent 6890N (Wilmington, DE) with dual micro ECD detectors, dual split/splitless inlets and an autosampler for simultaneous injection on the two capillary columns. The ^{63}Ni detectors have a small cell volume, use pulse frequency signal processing and have a wide linear dynamic range. An Agilent ChemStation data system was used for instrument control and data acquisition and processing. Inlets were operated in the splitless mode with a uniform injection volume of 2 µL. The primary GC column was a 0.32 mm i.d. X 30 m Agilent fused silica capillary column with an 0.25 µm HP-5 stationary phase. This bonded phase (or an equivalent chemically-bonded 5% phenyl-95% methyl stationary phase) is widely used for CB analysis as it provides excellent resolution and performs over a wide temperature range. The second (or confirmation) column used was a 30 m X 0.25 mm (i.d.) J & W Scientific DB-1701 with an 0.25 µm film thickness. Again, an equivalent chemically-bonded cyanopropylphenyl-methyl polysiloxane capillary column could be substituted.

The 6890N operating conditions were: injector, 250°C, purge time, 0.75 min, purge flow, 60 mL He/min; oven temperature program, 80°C for 2 min, 40°C to 175°C; 2°C/min to 300°C and hold 5 mm; run time, 71.4 min. The column was operated in constant pressure mode with 9 psig He head pressure for the 0.32 mm column and 18 psig for the 0.25 mm column. The ECD detectors were operated at 325°C with 30 mL/min 95% argon-5% methane as make up gas and a data acquisition rate of 5 Hz.

A Finnigan DSQ GC-MS was used to identify substances in the SSTs and cryovials. Reagent water or a water-hexane mixture (3:1, v/v) sat in the tubes for one hour. The hexane layer (or hexane extract of the water rinse) was dried over sodium sulfate and analyzed using gas chromatography conditions similar to those described above. Compounds were tentatively identified by a computerized library search of the NIST mass spectral library.

Serum Extraction and Extract Fractionation

Serum was thawed and allowed to reach room temperature. The available specimen, but not more than 2.0 mL, was transferred to tared glassware for determination of weight before adding 2 mL of methanol followed by vigorous agitation for a minimum of one min. n-Hexane:diethyl ether (6 mL, 1:1 v/v) was added and the mixture was shaken vigorously for 2 min before centrifugation (2000 g for 10 min). The organic layer (4 mL) was removed and exchanged to a small volume (~0.25 mL) of isooctane in a nitrogen evaporator. This extraction method is similar to that reported by Luotamo et al. (6) for determination of serum polychlorinated biphenyls.

The serum extract was dried with a miniature glass column containing ~0.5 g of hexane-washed sodium sulfate retained with a glass wool plug. A 2 cm (i.d.) glass chromatography column with a glass wool plug and a Teflon stopcock was dry packed with 5 g of silica gel – the adsorbent required overnight activation in a 130°C oven. The column was gently tapped to avoid spaces, topped with ~1 cm of anhydrous sodium sulfate and rinsed with 30 mL of n-hexane. A receiver was placed under the column and the extract transferred quantitatively in several 2 mL portions of n-hexane. After the sample percolated into the bed, a total of 35 mL of n-hexane was added to the column reservoir: the first 15 mL collected was fraction F_1 and the following 20 mL collected was fraction F_2. A further 10 mL of anhydrous benzene was collected as fraction F_3. Fractions F_2 and F_3 were exchanged to a small volume of isooctane in a nitrogen evaporator (6.6% of the serum volume is required for a 10-fold concentration factor) and the samples were transferred to autosampler vials. A similar column cleanup was used for analysis of organochlorine compounds in fish, human milk and other fatty samples (7). The activity of the silica gel was checked by verifying elution of CBs in F_2 and BDEs in F_3. 4,4'-DDT and the other DDT compounds elute predominantly in F_3, but 4,4'-DDE splits between F_2 and F_3.

During method development some extracts were defatted using the method of Luotamo et al. (6). The hexane:diethyl ether extract was concentrated to ~0.5 mL under nitrogen, adjusted to 2.0 mL with hexane and combined with 2 mL of sulfuric acid. After vigorous shaking the emulsion was broken by centrifugation for 10 min at 2,000 rpm before the organic layer was removed and dried with sodium sulfate.

The usual precautions for trace organic analysis of electron capturing compounds were followed. Glassware was scrupulously cleaned with detergent solutions, triple rinsed with deionized water, air dried and solvent washed before use. Glassware was stored in enclosed cabinets with foil covers. All solvents were reserved for the study and sample preparation was conducted in an isolated area where contact with plastics was minimized. Similar precautions have been noted in previous reports (8, 9).

In order to check for possible specimen contamination from the SSTs, a preliminary study was undertaken. Blood was drawn from volunteers and held in either solvent-rinsed glass tubes or SSTs. Blood in glass tubes was allowed to clot in the refrigerator for 2 – 4 hours prior to isolation of the serum by centrifugation.

Pilot Study

XAFP specimens were collected from Caucasian women in 2002 in three southern California counties. The specimens were selected randomly and identifiers linking them to subjects were removed. Two protocols were used to process specimens. Half of the sera (samples numbered 1 – 20) were frozen directly leaving the sera in contact with the gel barrier and the red blood cell pellet and hemolysis products during storage at –20°C. The remaining specimens (numbered 21 – 40) were decanted and frozen in the cryovials. The latter protocol was examined as a precautionary step to minimize artifacts and maintain serum integrity. The sera were otherwise subjected to all the manipulations and handling involved in routine prenatal screening.

SPECIAL PRECAUTIONS

Human blood and serum should be handled with Biosafety Level 2 precautions or higher as they are potentially infectious (10). Benzene is a known human carcinogen and should be handled only with appropriate protective clothing and in an efficient fume hood. Many of the target compounds are toxic and should only be handled with impermeable gloves, lab coats and face protection, especially when handling neat materials and concentrated solutions.

Results

Method Development and Preliminary Experiments

Our method development effort proceeded stepwise from very simple approaches (e.g., direct GC analysis of serum extracts) to more elaborate as required. Direct GC analysis of calf serum extracts was only effective for determination of 4,4'-DDE, the residue typically in highest abundance in the blood. Without supplemental cleanup the baselines were noisy and chromatograms had a pronounced hump on which 4,4'-DDE eluted as a shoulder. Defatting extracts with sulfuric acid eliminated the rise in the baseline and improved integration of DDT compounds, but acid treatment also resulted in a pronounced sag later in the chromatogram and introduced negative ECD peaks. The negative chromatographic peaks created problems in automated integration of the chromatograms.

SSTs introduced artifacts to the serum specimens. Water rinses of the SSTs contained a variety of hydrocarbons including normal, aliphatic hydrocarbons from C_{21} to C_{27}, antioxidants (butyrated hydroxytoluene) and a plasticizer (diisooctyl phthalate). These artifacts were identified by GC-mass spectrometry where the spectra of the unknowns matched the library spectra closely, and satisfied both forward and reverse matching criteria. The cryotubes also contained readily extracted substances that could interfere with trace analysis. Hexane rinses contained two prominent components, dodecylacrylate (major) and dodecanol or dodecylacetate. Electron ionization mass spectra of dodecanol and its acetate are not distinguishable.

Whole blood specimens processed in glass and analyzed by GC-ECD were reduced in early eluting compounds as well as two major artifacts from the SST tubes, one of particular importance was the phthalate plasticizer because of its large electron capture detector response. In contrast, other major artifacts present, particularly the hydrocarbons from the SST gel, are not detected by ECD-GC and don't cause interference. Similarly, the dodecylacrylate (from cryotubes) is detected by the ECD-GC introducing an additional artifact.

Interferences from the plastic sample tubes could be eliminated in either of two ways: by treatment of extracts with sulfuric acid or by silica gel column chromatography. When both of the cleanup techniques were applied (i.e., serum extracts were defatted then fractionated on silica gel), the highly retained artifacts depressing the baseline eluted in the discarded F_1 fraction, and both the F_2 (CB fraction) and F_3 (DDTs and BDEs) fractions had flat, easily-integrated baselines. The simpler approach, using only silica gel cleanup, produced the best chromatograms.

The 26 method analytes are summarized in Table 1 and include the chlorobiphenyl congeners in CLB-1 mixture D, the major components of the most common commercial Aroclors. In addition 4,4'-DDE, the major DDT metabolite, and 4,4'-DDD were included as well as their 2,4'-substituted analogues. The tetra, penta, and hexabrominated diphenyl ethers studied are abundant components of commercial BDE flame retardants (11). Among the 26 standards retention times (t_R) ranged from 12 min to 58 min on the two capillary columns. There are 4 difficult-to-resolve pairs on the primary GC column (2,4'-DDE/CB-101; 4,4'-DDD/2,4'-DDT; 4,4'-DDT/CB-138; BDE- 100/CB-194). Among these pairs all but the last reverse retention order on the confirmation column, and 2,4'-DDD and CB-151 also elute in reverse order.

Method Validation

Reagent water blanks were free of detectable DDTs as well as the tetra-, penta- or hexabromo BDEs. In sharp contrast, the chlorobiphenyls were routine laboratory contaminants. Laboratory analyses were conducted in a 50-year-old building contaminated with traces of Aroclors associated with fluorescent lighting, and possibly contaminated building materials (12). Our laboratory relocated to a new building in 2003 after completion of this work.

The human serum SRM has certified CB concentrations of 170 to 670 pg/mL with expanded uncertainties of 6 to 18% (RSD). Without background correction there was a systematic high bias for CBs with recoveries ranging from 108 to 183%. With blank correction the accuracy was improved, i.e., mean recoveries were $115 \pm 11\%$ and ranged from 96 to 123% (Table 2). The mean result for 4,4'-DDE was 7.68 mg/mL (RSD 16%) relative to a certified concentration of 6.6 mg/mL and accuracy was similar for 4,4'-DDT even though its SRM concentration was ~100-fold lower.

Table 1. Method Analytes, Abbreviations, CAS Numbers and Typical GC Retention Times

Analyte	Abbreviation	CAS No.	Retention Time (min)	
			HP-5	DB-1701
4,4'-Dichlorobiphenyl	CB-15	2050-68-2	12.52	13.99
1,1-Dichloro-2-(2-chlorophenyl)-2-(4-chlorophenyl)ethylene	2,4'-DDE	3424826	21.36	23.61
2,2',4,5,5'-Pentachlorobiphenyl	CB-101	37680-73-2	21.52	23.42
2,2-bis(p-chlorophenyl)-1,1-dichloroethylene	4,4'-DDE	72-55-9	23.56	26.05
1,1-Dichloro-2-(2-chlorophenyl)-2-(4-chlorophenyl)ethane	2,4'-DDD	53-19-0	24.15	28.38
2,2',3,5,5',6-Hexachlorobiphenyl	CB-151	52663-63-5	24.94	26.97
2,3',4,4',5-Pentachlorobiphenyl	CB-118	31508-00-6	25.94	28.59
1,1-bis(4-chlorophenyl)-2,2-dichloroethane	4,4,-DDD	72-54-8	26.62	32.53
1,1,1-Trichloro-2-(2-chlorophenyl)-2-(4-chlorophenyl)-ethane	2,4'-DDT	789026	26.83	29.54
2,2',4,4',5,5'-Hexachlorobiphenyl	CB-153	35065-27-1	27.66	29.83
2,2',3,4,5,5'-Hexachlorobiphenyl	CB-141	52712-04-6	28.62	31.18
1,1-bis((p-chlorophenyl)-2,2,2-trichloroethane	4,4'-DDT	50-29-3	29.49	33.52
2,2',3,4,4',5'-Hexachlorobiphenyl	CB-138	35065-28-2	29.78	32.52

2,2',3,4',5,5',6-Heptachlorobiphenyl	CB-187	52663-68-0	31.12	-[a]
2,2',3,4,4',5,5'-Heptachlorobiphenyl	CB-180	35065-29-3	35.24	-[a]
2,2',4,4'-Tetrabromodipheny ether	BDE-47	5436-43-1	35.39	39.19
2,2',3,3',4,4',5-Heptachlorobiphenyl	CB-170	35065-30-6	37.62	-[a]
2,2',3,3',4,5,5',6'-Octachlorobiphenyl	CB-199	52663-75-9	38.37	-[a]
2,2',3,3',4,4',5,6'-Octachlorobiphenyl	CB-196	42740-50-1	38.83	-[a]
2,2',3,3',4,4',5,6-Octachlorobiphenyl	CB-195	52663-78-2	41.24	-[a]
2,2',4,4',6-Pentabromodiphenyl ether	BDE-100	189084-64-8	42.75	46.15
2,2',3,3',4,4',5,5'-Octachlorobiphenyl	CB-194	35694-08-7	42.95	46.21
2,2',4,4',5-Pentabromodiphenyl ether	BDE-99	60348-60-9	44.87	48.74
Decachlorobiphenyl	CB-209	2051-24-3	48.56	49.90
2,2',4,4',5,6'-Hexabromodiphenyl ether	BDE-154	207122-15-4	50.88	54.26
2,2',4,4',5,5'-Hexabromodiphenyl ether	BDE-153	68631-49-2	53.78	57.78

[a]Individual standard available and retention order for 5% phenyl phase from reference

**Table 2. NIST SRM 1589a and Laboratory Reagent Water Blank
Analysis Results and Analyte Recoveries**

	Concentration (ng/mL)						
	4,4'-DDE	4,4'-DDT	CB-153	CB-138	CB-187	CB-180	CB-170
NIST Certified Concentration	6.60	0.09[a]	0.67	0.48	0.17	0.48	0.19[a]
NIST Expanded Uncertainty	1.00	0.01	0.04	0.04	0.03	0.03	0.00
SRM[b] #1	6.6	0.15	0.65	0.62	0.28	0.69	0.25
SRM #2	6.8	0.11	0.65	0.66	0.35	0.76	0.29
SRM #3	9.2	0.17	0.84	0.84	0.28	0.81	0.33
SRM #4	8.1	0.00	0.75	0.75	0.35	0.70	0.32
Mean (SD)	7.7 (1.2)	0.11 (0.08)	0.72 (0.09)	0.72 (0.10)	0.32 (0.04)	0.74 (0.06)	0.30 (0.04)
Recovery (%)	116	126	108	149	183	153	160
LRW[c] #1	0.00	0.00	0.07	0.12	0.10	0.12	0.06
LRW #2	0.16	0.00	0.15	0.24	0.18	0.25	0.11
LRW #3	0.00	0.00	0.02	0.04	0.06	0.07	0.05
Mean (SD)	0.05	0.00	0.08	0.13	0.11	0.15	0.08
Corrected Recovery (%)	115	126	96	121	118	123	119

[a]Reference concentration
[b]Standard reference material
[c]Laboratory reagent water

2,2',4,4'-Tetrabromodiphenyl ether (BDE-47) was determined in the reference serum at a concentration of 138 ± 44 pg/mL, but other BDEs were below the method detection limit. Certified or reference data for the BDEs in the reference material have not been published and we were otherwise unable to identify a reference serum for these pollutants.

These SRM data demonstrate that the serum analysis method developed is accurate for determination of three types of environmental contaminant studied, but only when the CB residues are in high parts-per-trillion concentrations (or the laboratory is cleaner). Because of the contamination in our former laboratory, we could not reliably determine the low concentrations of CBs in the XAFP specimens.

XAFP Pilot Study

DDT Residues

4,4'-DDE was detected in each XAFP serum specimen analyzed. DDE ranged from a low of 170 pg/mL to a high of 8,900 pg/mL, yet only one blood specimen exceeded 2 mg/mL as seen in the frequency distribution in Figure 1. The arithmetic mean was 1.00 ± 1.52 pg DDE/mL with the high specimen exceeding this mean by >5 standard errors. We found no evidence that storing serum over the red blood cells in the SSTs affected the results, i.e., the observed and expected distributions for the two groups were about the same.

Only 31 XAFP specimens were evaluated in the final data set as 9 of the samples were consumed by analysis using a preliminary and less effective version of the method. Midway through the study we found that a higher concentration factor was required for determination of the pollutants at the low concentrations present. In contrast to DDE, 4,4'-DDT was detected in only two (or 13%) of the specimens. The DDE:DDT ratio was ~20:1 in the high s-DDT (sum of 6 DDTs) sample.

DDE is intermediate in polarity eluting in both F_2 (~57%) and F_3 (43%). The residue in both fractions must be summed for accuracy. There was good quantitative agreement between the two GC columns, i.e., F_2 (HP-5, 4.99 mg/mL; DB-1701, 5.06 mg/mL) and F_3 (HP-5, 3.76 ng/mL; DB-1701, 3.88 mg/mL). Chromatograms of extracts fractionated on silica gel were largely free of interferences and baseline irregularities unlike the crude or sulfuric acid-defatted extracts described above. Typical XAFP chromatograms for F_2 and F_3 are shown in Figures 2 and 3, respectively.

BDE Residues

BDEs in the maternal sera showed greater specimen-to-specimen variation (Table 3). Tetrabromodiphenyl ether (BDE-47) was the most widely distributed BDE congener, exceeding the detection limit in 55% of samples. The penta- and hexabromodiphenyl ethers, by contrast, were detected in only 7 to 10% of the specimens. BDE flame retardants, like DDE residues, appeared to have a bimodal distribution where a small proportion of subjects have relatively high body burdens. The arithmetic mean of s-BDE was 0.31 ± 0.45 mg/mL -- two subjects had s-BDE levels 2.4 and 3.5 standard errors higher (specimens with no detectable s-BDE were rejected from the sample statistics). While the two high s-BDE specimens were among those stored with the pellet, there was no evidence of systematic BDE contamination. The observed and expected distributions are comparable in both groups. About half of the specimens with no detectable s-BDE were from the group stored with the blood pellet in SSTs (56%) and about half were stored in cryovials (44%).

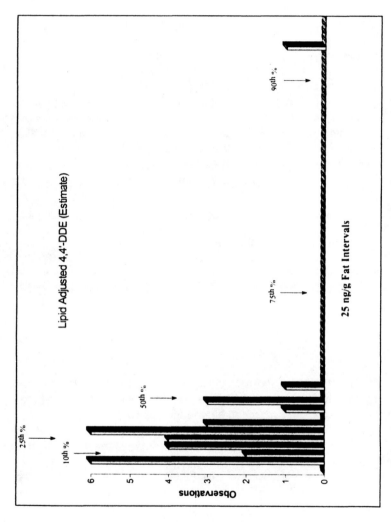

Figure 1. Frequency distribution of serum 4,4'-DDE levels in XAFP maternal sera. Percentiles represent U.S. female population (adapted from reference 21).

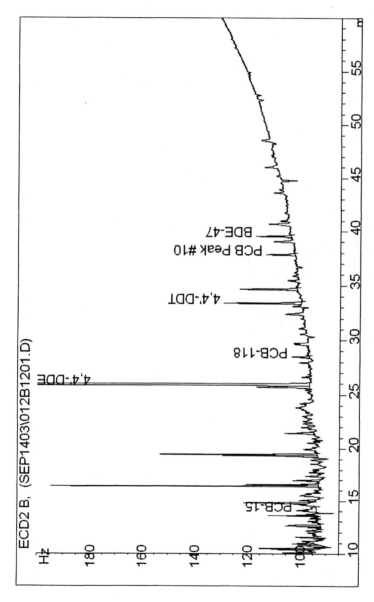

Figure 2. XAFP serum fraction F₂ containing analyzed on the DB-1701 capillary column.

62

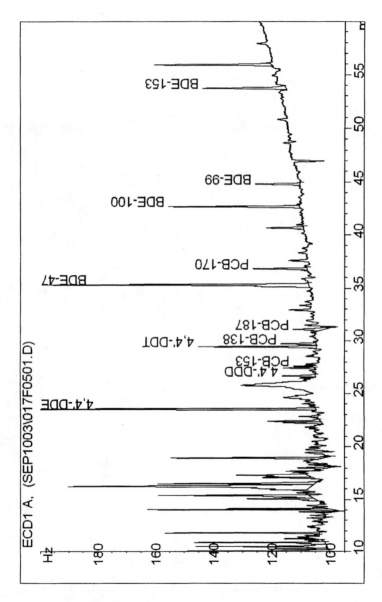

Figure 3. XAFP serum specimen #16 fraction F₃ containing 4,4'-DDE and four prominent BDE congeners.

Table 3. Polybrominated Diphenyl Ethers in Banked Maternal Serum
XAFP Specimens[a]

Specimen	BDE Concentration (ng/mL)				
	BDE-47	BDE-100	BDE-99	BDE-153	s-BDE
16	1.0	0.18	0.48	0.21	1.87
6	0.82	0.30	ND	0.26	1.38
10	0.49	ND	ND	ND	0.49
23	0.47	ND	ND	ND	0.47
33	0.084	ND	ND	0.16	0.24
39	0.18	ND	ND	ND	0.18
37	0.16	ND	ND	ND	0.16
18	0.14	ND	ND	ND	0.14
19	0.13	ND	ND	ND	0.13
13	0.12	ND	ND	ND	0.12
21	0.12	ND	ND	ND	0.12
24	0.12	ND	ND	ND	0.12
29	0.12	ND	ND	ND	0.12
36	0.11	ND	ND	ND	0.11
30	0.10	ND	ND	ND	0.10
34	0.086	ND	ND	ND	0.086
38	0.074	ND	ND	ND	0.074

[a]Fourteen additional specimens had no detectable s-BDE

ND = Not detected

Specimens with the highest s-BDE had 3 or 4 BDE congeners including BDE-47, BDE-100 and BDE-153. BDE-99, the congener with the lowest electron capture detector response, was found in only one specimen shown in Figure 3. The silica gel F_3 chromatograms again had flat, reliably integrated baselines. In the serum shown BDEs are among the most prominent peaks with areas similar to the 4,4'-DDE residue. There was good quantitative agreement for results from the two capillary columns, e.g., BDE-47 (HP-5, 0.99 mg/mL and DB-1701, 1.0 ng/mL); BDE-99 (HP-5, 0.44 mg/mL and DB-1701, 0.52 ng/mL); BDE-100 (HP-5, 0.15 mg/mL and DB-1701, 0.20 ng/mL); and BDE-153 (HP-5, 0.21 mg/mL and DB-1701, 0.21 ng/mL).

Discussion

GC analysis of organochlorine pesticides in human serum was first described almost 40 years ago (*13*). Since that report a number of modifications of the original "triple hexane" liquid-liquid extraction method have appeared, some with improved recoveries of lipophilic pollutants (*6, 14, 15*). These modifications involved different extraction solvents and mixed solvents, use of different approaches to break emulsions, and new target compound lists. Bouwman et al. (*16*) avoided emulsions by immobilizing serum on diatomaceous earth mixed with silica, and eluted pesticides with organic solvent. Denaturing serum proteins either with methanol (*6*) or formic acid (*8*) improved extraction efficiencies. Sample cleanup, either by treatment with sulfuric acid (*6*) or adsorption chromatography on silica, alumina, or Carbopak (*17*) gave cleaner chromatograms and lower detection limits. Solid phase extraction (SPE) methods for determination of pesticides (*18*) and pesticides and PCBs in serum (*8, 19*) also have been reported.

The method described here is similar to many of these earlier approaches. We chose to use liquid-liquid extraction, because it appears to give higher and more reproducible recoveries. We also found that the defatting step could be eliminated for these small samples, and that satisfactory cleanup could be achieved with column chromatography alone. The present method is rapid, yet has similar PCB and BDE detection limits to more elaborate and time consuming procedures (*6, 20*). The estimated sample throughput is up to an order of magnitude higher than high resolution MS methods for organohalogen pollutants greatly lowering analysis costs.

Dual capillary column analysis is reliable because contaminants that may coelute on a single high efficiency GC column are usually resolved on a second capillary. In this case the 5% phenyl phase separates primarily by a dispersive mechanism, while the cyanopropylphenyl column has polar and polarizable interactions as well. The dual column technique provides rapid confirmation and, because the instrument performs simultaneous two-column analysis, it is more efficient than sequential confirmation (*9*). Dual column confirmation is well established in regulatory analysis, and Brock and coworkers (*8*) previously used the technique for determination of pesticides and PCBs in serum.

A central question of this investigation was whether POPs could be reliably determined in XAFP serum specimens in spite of the widespread occurrence of these pollutants. In particular, the potential for specimen contamination was great because the specimens are collected in plastic tubes, and no special precautions were taken by the phlebotomists other than the usual ones mandated for safety. The data on DDTs and BDEs provide no evidence of specimen contamination or artifacts. CB determination, however, was judged to be unreliable because of laboratory contamination, even though the method was accurate based on SRM analyses. We are further investigating whether CB contamination is reduced in a specially designed "organics free" laboratory at our new building. The question of low level CB contamination by plastic SSTs and cryovials also needs to be examined further.

There is a considerable database on 4,4'-DDE serum levels in the U. S. population and this database provides an important reference for the levels in the XAFP specimens (*21*). The NHANES U.S. population data for the 1999-2000 survey on a sample of 1,027 females are shown as the indicated 10th, 25th, 50th, 75th and 90th percentile in Figure 1. We did not measure fat content in our specimens, but normalized the data by assuming an average serum fat content of

0.6% (w/w). The frequency distribution has 25 mg/g increments or classes. If the two populations were the same, 50% of the XAFP specimens would fall below the plotted 50th percentile, etc. Our population may be different as we studied young, reproductive age women and POPs body burdens tend to increase with age. There are also influences of race and ethnicity and we studied Caucasian women. Thus, in addition to geographic differences, the age of these women may be a factor as well as the ethnic makeup of the sample. There were no observations in the lowest interval because it was below the detection limit. The single high value cannot be considered an outlier as it is between the 90th and 95th percentile for the U.S. female population.

There is far less information on BDEs. We compared the BDE congener distributions for specimen # 16 with maternal serum from Indiana women (*22*) and year 2000-2002 serum pools from U. S. cities (*23*). As seem in Figure 4 the patterns are striking in their similarity when one considers the geographic differences and unique population characteristics (in the case of the serum pool).

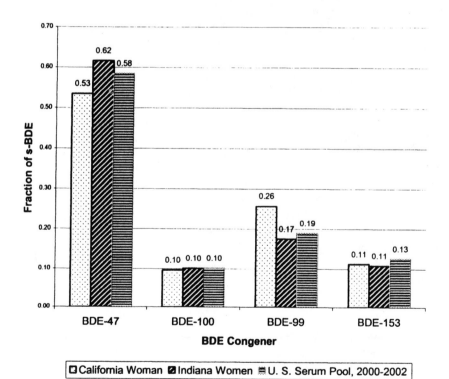

Figure 4. BDE congener profiles for California XAFP serum #16, material serum from Indiana study (median, n=12) and U.S. serum pools collected in 2000-2002 (median). For consistency, s-CB was defined as the 4 BDE congeners plotted.

BDE residues in the California woman appear to be typical of subjects from Indiana, Louisville, Miami, Memphis, and Philadelphia. Perhaps the elevated BDE body burdens in two women in our limited survey are associated with occupational exposures in printed circuit board assembly, or work with polyurethane foam products (20). The correlates of BDE exposure among California women are of considerable interest for further study.

Serum provides a good estimate of the total body burden of organohalogen pollutants, especially when corrected for serum lipids. Analysis of paired adipose and serum specimens for mirex, for example, established correlation coefficients of 0.818, 0.847, and 0.838 on a whole serum basis, lipid-adjusted serum basis, and serum albumin-adjusted basis, respectively (14). There also is a strong relationship between maternal blood and lipophilic pollutants in breast milk with the fat-adjusted milk:blood ratio close to one. The absolute concentration of lipophilic pollutants in mother's milk is higher as human milk has a much higher lipid content, up to ~3.5% by weight. Individual fetal blood concentrations of BDE do not differ significantly from maternal blood (22). Another recent report confirms that the partition of BDE between adipose and milk or blood lipids is close to unity (24). Thus, maternal blood pollutant levels are closely tied to both prenatal exposures and exposures in breast-fed infants.

The human health effects of DDT were recently reviewed by Longnecker and colleagues (25). DDT is linked to premature births (26) and prematurity represents a major cause of infant mortality in the U.S. DDT levels occurring in the majority population at this time are only about 1/5th of levels in the 1960's, therefore it is believed that an effect on prematurity was more probable in that era. Many of the epidemiologic studies of ambient DDT exposures and associations to breast and other cancers, impairment of lactation, neurological or developmental abnormalities have been either negative or inconclusive. Low-level DDE exposure may impact the endocrine system and has been shown to antagonize androgen. Subtle effects of DDT on human health may be confounded by obesity in the general population. For example, obesity is associated with higher serum DDE levels. In the occupational environment DDT neurotoxicity (DDT poisoning) has been observed, and DDT has been associated with pancreatic cancer, non-Hodgkin's lymphoma and changes in blood liver function enzymes.

BDEs are of concern in humans because of their associated developmental neurotoxicity, effects on thyroid hormones, and reproductive toxicity, the most sensitive effects recognized to date (27). In experimental animals BDEs are hepatotoxic, induce enzyme systems and are hepatocarcinogens at high doses. Due to their toxicity, elevated BDE levels in pregnant women are believed to represent a health risk to both the embryo and nursing newborn (28).

Conclusions

This study demonstrates that discarded XAFP specimens can yield exposure information on important environmental contaminants in pregnant women. While the SSTs used in XAFP sampling (as well as the cryotubes proposed for storage) introduce artifacts including antioxidants and plasticizers, simple cleanup techniques such as silica gel column chromatography are effective for removing them. Other artifacts such as the saturated hydrocarbons are not detected by the electron capture detector. In the cases of the BDEs and DDTs, we find no evidence of a need for special handling of XAFP specimens. The situation with CBs, however, is not clear and requires further study.

These findings are important in public health research for several reasons. First, there is a paucity of biomonitoring-based chemical exposure information, in part, because large numbers of human specimens are so difficult to obtain. Use of discarded XAFP samples is an efficient means to gain access to a large cross section of the population, without significant outlay of additional resources or added discomfort or risk to human subjects. Analysis of these specimens could give valuable information on reference ranges in the regional population as well as constituent groups. There is growing interest in public health tracking where chemical exposure data is fundamental to assess status and trends (29). Biomonitoring data also can provide a sentinel function for early detection of emerging pollutants (3).

Finally, XAFP specimens can be linked to existing health data such as records on birth outcomes, birth defects and a variety of diseases such as autism, childhood leukemia and asthma. The ability to link authoritative chemical exposure information to disease registries addresses a fundamental hurdle of environmental epidemiology.

Acknowledgement

Donald Wijekoon, Michael McKinney and Paramjit Behniwal (California Department of Health Services) and Christine Arneson, Thu Quach and Lori Copan (California Biomonitoring Planning Project) provided assistance. Myrto Petreas (California Department of Toxic Substances Control) provided information and discussions on occurrence and analysis of PBDE. Drs. Larry Needham (U.S. Centers for Disease Control and Prevention) and Marita Luotamo (Finnish Institute of Occupational Health) contributed information on serum analysis for POPs. We thank Michelle Marzullo, Michelle Pearl and George Cunningham, MD (CDHS Genetic Disease Branch) for assistance with identification and accession of blood specimens. We thank the U.S. Department of Health and Human Services, Centers for Disease Control and Prevention (CDC) for support under Program Announcements 01072 and Cooperative Agreement U90/CCU917016-02.

References

1. Kaufman, F. L.; Kharrazi, M.; Delorenze, G.; Eskenazi, B.; Bernert, J. T. *J. Exp. Anal. Environ. Epidemiol.* **2002**, *12*, 286-295.
2. Goldman, L. R.; Anton-Culver, H.; Kharrazi, M.; Blake, E. *Environ. Health Perspect.* **1995**, *103 (Suppl)*, 31-34.
3. Lee, L. W.; Griffith, J.; Zenick, H.; Hulka, B. S. *Environ. Health Perspect.* **1995**, *103 (Suppl 3)*, 3-8.
4. Smith, D. *Intern. J. of Epidemiol.* **1999**, *28*, 179-188.
5. Petreas, P.; She, J.; Brown, F. R.; Winkler, J.; Windham, G.; Rogers, E.; Zhao, G.; Bhatia, R.; Charles, M. J. *Environ. Health Perspect.* **2003**, *111*, 1175-1179.
6. Luotamo, M.; Jarvisalo, J.; Aitio, A. *Environ. Health Perspect.* **1985**, *60*, 327-332.
7. Draper, W. M.; Koszdin, S. *J. Agric. Food Chem.* **1991**, *39*, 1457-1467.
8. Brock, J. W.; Burse, V.W.; Ashley, D.L.; Najam, A. R.; Green, V.E.; Korver, M.P.; Powell, M.K.; Hodge, C. C.; Needham, L. L. *J. Anal. Toxicol.* **1996**, *20*, 528-536.
9. Greizerstein, H. B.; Gigliotti, P.; Vena, J.; Freudenheim, J.; Kostyniak, P. J. *J. Anal. Toxicol.* **1997**, *21*, 558-566.
10. *Biosafety in Microbiological and Biomedical Laboratories*, U.S. DHHS, Washington, DC: Centers for Disease Control and Prevention/National Institutes of Health, 1998.
11. Peltola, J.; Yla-Mononen, L. *Pentabromodiphenyl ether as a global POP*; Finnish Environment Institute, Chemicals Division, TemaNord, 2000, URL http://www.unece.org/env/popsxg/pentabromodiphenyl_ether.pdf
12. Herrick, R. F.; McClean, M. D.; Meeker, J. D.; Baxter, L. K.; Weymouth, G. A. *Environ. Health Perspect.* **2004**, *112*, 1051-1053.
13. Dale, W. E.; Curley, A.; Cueto, C. *Life Sci.*, **1966**, *5*, 47-54.
14. Burse, V. W.; Head, S. L.; McClure, P. C.; Korver, M. P.; Alley, C. C.; Phillips, D. L.; Needham, L. L.; Rowley, D. L.; Kahn, S. E. *J. Agric. Food Chem.* **1989**, *37*, 692-699.
15. Greve, P. A.; Van Zoonen, P. *Intern. J. Environ. Anal. Chem.* **1990**, *38*, 265-277.
16. Bouwman, H; Sydenham, E.W.; Schutt, C. H. J. *Chemosphere* **1989**, *18*, 2085-2091.
17. Gill, U.S.; Schwartz, H. M.; Wheatly, B. *Chemosphere* **1995**, *30*, 1969-1977.

18. Saady, J. J.; Pooklis, A. *J. Anal. Toxicol.* **1990**, *14*, 301-304.
19. Pauwels, A.; Wells, D. A.; Covaci, A.; Schepens, P. J. C. *J.Chromatogr. B* **1999**, *723*, 117-125.
20. Sjodin, A.; Hagmar, L.; Klasson-Wehler, E.; Kronholm-Diab, K.; Jakobson, E.; Bergman, A. *Environ. Health Perspect.* **1999**, *107*, 643-648.
21. *Second National Report on Human Exposure to Environmental Chemicals,* Department of Health and Human Services, Centers for Disease Control and Prevention, January, 2003, page 190-193.
22. Mazdai, A.; Dodder, N. G.; Abernathy, M. P.; Hites, R. A.; Bigsby, R. M. *Environ. Health Perspect.* **2003**, *111*, 1249-1252.
23. Sjodin A.; Jones, R. S.; Lapeza, C.; Focant, J. F-.; Wang, R.; Turner, W. E.; Needham, L. L.; Patterson, D. J., Jr. *Organohalogen Compounds* **2003**, *61*, 1-4.
24. Meironyte, G. D.; Aronsson, A.; Ekman-Ordeberg, G.; Bergman, A.; Noren, K. *Environ. Health Perspect.* **2003**, *111*, 1235-1241.
25. Longnecker, M. P.; Rogan, W. J.; Lucier, G. *Annu. Rev. Public Health* **1997**, *18*, 211-244.
26. Longnecker, M. P.; Klebanoff, M. A.; Brock, J. W.; Zhou, H.; Gray, K. A.; Needham, L. L.; Wilcox, A. J. *Am. J. Epidemiol.* **2002**, *155*, 313-322.
27. Birmbaum, L. S.; Staskal, D. F. *Environ. Health Perspect.* **2004**, *112*, 9-17.
28. Schecter, A.; Pavuk, M.; Papke, O.; Ryan, J. J.; Birnbaum L.; Rosen R. *Environ. Health Perspect.* **2003**, *111*, 1723-1729.
29. *California Biomonitoring Plan,* State of California Department of Health Services, 2003, URL http://www.catracking.com/sub/new.htm

Chapter 5

Assessing Exposure to Agricultural Fumigants in Outdoor and Indoor Air Environments

James E. Woodrow[1] and Robert I. Krieger[2]

[1]Natural Resources and Environmental Science, University of Nevada, Reno, NV 89557
[2]Department of Entomology, Personal Chemical Exposure Program, University of California, Riverside, CA 92521

Because of ongoing concerns over exposure to agricultural fumigants, air sampling field measurement methods and computer-based models, which use field and chemical property data as input, have been developed for determining fumigant volatilization losses from target sites and subsequent downwind concentrations in non-target areas for exposure assessment. Air sampling methods have shown that current application practices lead to substantial emission losses of many of the common soil fumigants. As a result, human inhalation exposures – outdoors and indoors – often exceed toxicological acute and sub-chronic reference concentrations, indicating a potential health risk. In this regard, some computer-based models, using emission rates and meteorology, are used to establish safe buffer zones around fumigant sources. However, fumigant emissions, and thereby risk, could be lessened by containment of the fumigant in the soil column (impermeable film, water seal) or by chemical reaction in the soil to less harmful products (thiosulfate-containing fertilizer with the halogenated fumigants).

Introduction

With the use of agricultural chemicals to control pests for increased production comes the responsibility of, firstly, recognizing that non-target entities (i.e., humans, animals, ecosystems) will experience some unintended exposures to the chemicals and, secondly, assessing the magnitude of those exposures. As a class, soil fumigants are primarily characterized by high volatility, which enables them to penetrate and diffuse through soils for pest control. Compared to semi-volatile pesticides, which are commonly applied to soil surfaces, soil fumigants, which are often applied by injection at some soil depth (e.g., 10-30 cm), show about 400-8,000-times greater emission rates under typical field conditions, due primarily to their much greater vapor pressures (Table I). In agricultural regions of California where fumigants are commonly used with semi-volatile pesticides, a high priority is placed on reducing fumigant emissions because they make up the majority of the pesticide emissions inventory for most non-attainment areas in California (1).

Table I. Properties of common soil fumigants.

Fumigant	Structure	Vapor Pressure (Pa)	Sw^a (g/L)	Koc^b (mL/g)	Average Ef/Ep^c
Methyl Bromide	CH_3Br	1.9×10^5	13.4	83	6×10^3
Chloropicrin	Cl_3CNO_2	2,266	2.27	7	8×10^3
MITC[d]	$CH_3N{=}C{=}S$	2,533	7.60	32	400
1,3-D[e]	$ClCH_2CH{=}CHCl$	3,733	2.25	32	600
Carbon Disulfide	$S{=}C{=}S$	4.8×10^4	2.30	292	2×10^3

[a]Water solubility.
[b]Soil adsorption coefficient.
[c]E = emission rate ($\mu g/m^2 \cdot sec$), where Ef is for fumigant and Ep is for 11 semi-volatile pesticides applied to soil.
[d]Methyl isothiocyanate.
[e]Telone.

Since the atmosphere acts as the major transport medium for volatilized fumigants, the primary route of unintended exposure for humans and animals will be by inhalation. Fumigant toxicology studies have led to estimated inhalation reference concentrations (RfCs) for acute, sub-chronic, and chronic exposures (Table II). By definition, an RfC represents the level 'at or below which adverse non-cancer health effects are not estimated to occur' (2). Much of the discussion concerning exposures to soil fumigants in this chapter will focus on acute and sub-chronic exposures, with an emphasis on the latter. The

Table II. Fumigant inhalation reference concentrations.[a]

| | Reference Concentrations, $\mu g/m^3$ | | |
Fumigant	Acute (1-24 hours)	Sub-chronic (≥15 days)	Chronic (>1 year)
Methyl Bromide	815	8 (6 weeks)	5
Chloropicrin	29	1	1
1,3-D	109	14	9
MITC	66 (1-8 hours)	3	0.3
Carbon Disulfide	--	--	700[b]

[a] *2.*

[b] *34.* Baseline: 55,100 $\mu g/m^3$ (8-hour time-weighted-average [TWA]).

supporting data will be derived from a number of field studies concerned with exposures to agricultural soil fumigants. Descriptions will be made of the various measurement and modeling techniques that have been used and can be used for exposure assessment. But, modeling techniques will only be briefly described as a potential alternative to field measurement. Finally, the focus of this paper is on soil fumigants – exposures due to structural and commodity fumigations are not included in the discussion.

Field Measurement Methods

We are faced with the fact of fumigant emissions from treated soil, with movement of the vapors to non-target areas, and the task of assessing exposure to fumigant emissions, with their potential acute and sub-chronic human and ecological health impacts. Some reasonable approaches that have been made toward quantifying exposure include field measurements of exposure and estimation of exposure from measured and modeled emission rates. The latter will be described briefly below. Field methods that have been used to estimate emission rates of fumigants from soil include the following:

1. Aerodynamic gradient method (*3-5*)
$$ER = k^2 \Delta c \Delta u / (\Phi_m \Phi_p [Ln(z_2/z_1)]^2)$$
2. Integrated horizontal flux (*5,6*)
$$ER = (1/X) \int c_i u_i dz$$
3. Flux chambers (*7-10*)
$$ER = (V/A)(\Delta c/\Delta t)$$
4. Back calculation (*11*)
Downwind concentrations + ISC-ST model

In the above, 'ER' is emission rate ($\mu g/m^2 \cdot sec$), 'c' and 'u' are air concentration ($\mu g/m^3$) and wind speed (m/sec), respectively, 'z' is height (m), 'k' is the dimensionless von Kármán's constant (~0.4), 'Φ_m' and 'Φ_p' are atmospheric stability functions, 'X' is the depth of the source (m), 'V' and 'A' are the volume (m³) and area (m²), respectively, enclosed by the chamber, 't' is time (sec), and 'ISC-ST' is the industrial source complex-short term numeric atmospheric dispersion model (*12,13*). Briefly, 1) The aerodynamic method prescribes that the air sampling and wind speed masts be located at the center of the source, where the distance to the upwind edge is about 100 times the height of the masts (fetch); 2) For integrated horizontal flux, the sampling and wind speed masts are at the downwind edge of the source and their height is within the fumigant plume – height of the plume is about 10% of the depth of the source (X), depending on the stability of the atmosphere (*14*); 3) The use of flux chambers commonly involves determining the time rate of change of the fumigant concentration in air (*15*); 4) An iterative method is used with the ISC-ST model to achieve the best fit of the estimated emission rate with the observed downwind fumigant concentrations. Specifically, the field size, meteorology, terrain data, and an assumed emission rate are used as input to the ISC-ST model. Then, the downwind air concentrations simulated by the model using the assumed emission rate are statistically compared to the measured air concentrations. The result of this comparison is an adjustment or calibration factor for the assumed emission rate that will give the best fit with the observed downwind air concentrations.

Depending on the field method used, sampling media commonly include solid adsorbents (e.g., charcoal, polymers) and evacuated canisters that have had the interior surface specially treated to minimize chemical interaction with the air sample (*16*). The former is a cumulative method that traps and concentrates the fumigant, while canisters are 'whole air' samplers – composition of the sample is the same as that of the bulk atmosphere from which it is taken. The field data for methyl isothiocyante (MITC) and methyl bromide reported and discussed below were obtained using both coconut-based and petroleum-derived charcoal. An *in situ* method that shows promise for real-time determination of fumigants in the field is Fourier transform infrared (FTIR) spectroscopy. However, compared with the more commonly used cumulative and canister methods, the FTIR method is currently limited with regard to sensitivity (~0.2 ppm) (*17*).

Modeling Estimation Methods

Exposure can be estimated from the measurement of emission rates (field and laboratory) and the estimation of emission rates from models using field and laboratory data. Models have enjoyed increasing popularity in recent years, especially among government regulatory agencies, as computation power and reliability have improved. Some models have become successful rivals to field measurements, which have become costlier in terms of time and funding. Also, modeling can help the investigator conceptualize fumigant emission and atmospheric dispersion processes and to make sense of the field data. Modeling approaches that have been used to estimate emission rate and downwind dispersion include the following:

1. Emission Factor Documentation for AP-42, Section 9.2.2, Pesticide Application (*18*). This approach relates an emission factor – ratio of mass of pesticide volatilized to total mass applied – with pesticide vapor pressure. Emission rate is normalized to a 30-day period.

2. Pesticide Properties/Emissions Correlations (*19,20*). This approach related measured emission rates with the physicochemical properties of pesticides: LnER = mLnR + b, where 'ER' is emission rate and 'R' = (VP x AR)/(Sw x Koc x d). 'VP' is vapor pressure (Pa), 'AR' is application rate (kg/ha), 'Sw' is water solubility (mg/L), 'Koc' is soil adsorption coefficient (mL/g), and 'd' is depth of application (cm). The method is useful for estimating emission rates for new application scenarios and for new fumigants whose physicochemical properties are known.

3. Fumigant Emission Modeling System (FEMS) (*21*). This approach starts with measured emission rates – in this case for methyl isothiocyanate (MITC) emissions – and meteorology and uses Monte Carlo statistics to account for the uncertainty in emission rates and meteorology, with hourly updates over a several-day period. The Monte Carlo results are then used in the ISC-ST model to calculate concentration endpoints for various distances downwind for several years of simulations. These results can be related to level of exposure and defining a safe buffer.

4. Probabilistic Exposure and Risk Model for Fumigants (PERFUM) (*22*). This model uses historical meteorological data sets and iodomethane emission rates – estimated by back calculation using the ISC-ST model with measured 360° downwind concentrations as the input (*11*) – to calculate 360° downwind concentrations for buffer zones and margins of exposure for risk assessment for every day over a five-year period. The goal is to establish safe exposure distances for various application scenarios.

5. Support Center for Regulatory Air Models (SCRAM) (*12*). This is a U.S. EPA web site that contains a collection of computer-based atmospheric dispersion and receptor models for most exposure contingencies. The models of choice for the subject of this chapter are ISC-ST and AERMOD (AMS/EPA Regulatory Model) – multiple-source, complex terrain models that have receptor subroutines for risk assessment – and SCREEN – a single source, simple terrain model that uses the same algorithms as the ISC-ST. Compared to the other models, the SCREEN model is much easier to use and, as a screening tool, it can quickly give reliable order-of-magnitude levels of exposure, but it is somewhat limited in application (e.g., it is unable to handle complex terrains, complex meteorology, and multiple sources).

6. Lakes Environmental (*13*). This is a Canadian company that has available complete modeling packages that contain many of the EPA models, each with a graphical user interface for ease of data entry and operation.

Field Results

Solid adsorbent (e.g., charcoal) air sampling was used to monitor a number of fumigant applications and, in some cases, significant emissions occurred, depending on the application method (Table III). For example, methyl bromide exhibited an almost 90% loss over a 5-6 day period with no tarp (emission rate of ~370 $\mu g/m^2{\cdot}sec$) compared to up to 32% loss with a 1-mil high density polyethylene (HDPE) tarp (emission rate ~76-91 $\mu g/m^2{\cdot}sec$ for three different applications). Furthermore, when metam-sodium (Vapam) was applied at a 10 cm depth through a drip irrigation system, emission losses of MITC were minimal (3.5% [48 hrs]; emission rate ~5 $\mu g/m^2{\cdot}sec$). By contrast, application of metam-sodium by surface chemigation (a widely used method) led to MITC emission rates in excess of 300 $\mu g/m^2{\cdot}sec$, implying that cumulative losses were significantly greater as well. The data summarized in Table III suggest that current fumigant application practices will lead to airborne exposures in excess of the RfCs listed in Table II. For example, surface chemigation application of metam-sodium led to MITC air concentrations of 4,000-7,000 $\mu g/m^3$ at downwind distances of 5-150 meters (*23*). Compare this to 66 $\mu g/m^3$ for acute

Table III. Fumigant applications and subsequent losses.

Fumigant	Formulation	Application Rate (kg/ha)	Injection Depth (cm)[a]	Emission Rate ($\mu g/m^2{\cdot}sec$)	Cumulative Losses
Methyl Bromide	67/33[b]	263	25-30 (T)	76-91	22-32%
Methyl Bromide	67/33	263	25-30 (NT)	370	89%
Chloropicrin	99+	390	27 (T)	58-211	37-63%
Chloropicrin	99+	196	32 (NT)	180	62%
1,3-D	97	139	38 (NT)	9.7	12%
MITC	Metam-sodium	143[c]	10 (T, NT)	5	3.5% (48 hrs)
MITC	Metam-sodium	267[d]	Surf. Chem. Injection	>300	--[e]

[a]T = tarped (HDPE); NT = non-tarped.
[b]methyl bromide/chloropicrin.
[c]Equivalent to 81 kg/ha MITC.
[d]Equivalent to 151 kg/ha MITC.
[e]MITC concentration in air ≈ 4-7 mg/m^3 for surface chemigation (5-150 meters downwind [*23*]).

exposure (1-8 hrs.[Table II]). Even when application was done at a 10 cm depth by drip irrigation, MITC concentrations in air exceeded the acute RfC (up to about 3 times) for at least the first 6-8 hours (2-4 hours for application, 4 hours for next sampling period), depending on the downwind distance (Table IV).

Also for a methyl bromide application, concentrations in air were somewhat greater than the acute RfC (815 $\mu g/m^3$) up to 100 meters downwind during the application period.

Table IV. Fumigant concentrations in air associated with specific applications.

	Methyl Isothiocyanate (MITC)		Methyl Bromide		
A/SC/C[a]	Period (hours)	Concentration (μg/m³)	Period (days)	Concentration (μg/m³)	A/SC/C[a]
66/3/0.3	Application[b]	9-200[c]	Application	1,500[d]	815/8/5
	4	32-66	1	4-200[e]	
	8	27-51	2	3-60	
	24	22-39	3	3-22	
	36	11-18	4	5-7	
	48	14-18	5	0.5-15	

[a]A = acute, SC = sub-chronic, C = chronic exposures in $\mu g/m^3$.
[b]Drip irrigation, 10 cm.
[c]8-50 meters downwind. Compare to 4-7 mg/m³ at 5-150 meters for surface chemigation.
[d]6-100 meters downwind.
[e]6-611 meters downwind.

Similar occurrences were observed for air monitoring of ambient methyl bromide and MITC. Ambient methyl bromide was determined at a number of sampling stations distributed along the Salinas Valley, CA, during a typical application season (*24*). At two of the stations located within a few km of the treated fields (~33 applications), some nighttime levels exceeded the sub-chronic RfC (8 $\mu g/m^3$) (Figure 1). This is probably a fairly common occurrence. For ambient MITC – measured in an application area near Bakersfield, CA, during summer and winter months – the sub-chronic RfC (3 $\mu g/m^3$) was exceeded by a factor of about 2.0-5.4 for both indoor and outdoor stations during June and July, two of the heaviest application months (Figure 2) (*25*).

Indoor vs. Outdoor

The discussion thus far regarding exposure to fumigants has dealt primarily with outdoor air concentrations. The data summarized in Figure 2 show that indoor air concentrations of MITC were comparable to outdoor concentrations, and in some cases indoor levels even exceeded outdoor levels. Various studies have shown that outdoor airborne chemicals and particulates will infiltrate buildings (25-30). So, it is obvious that indoor concentration, C_i, is directly related to outdoor concentration, C_o, unless there are indoor sources, but C_i is

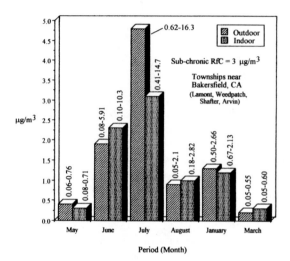

Figure 1. Ambient air concentrations (μg/m³) of methyl bromide in the Salinas Valley, CA (~33 applications).

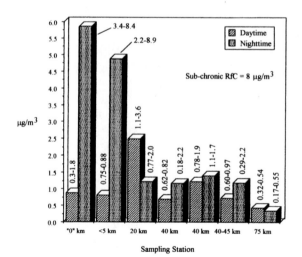

Figure 2. Ambient air concentrations (μg/m³) of methyl isothiocyanate (MITC) at townships near Bakersfield, CA (May-August: 30 metam-sodium applications; January-March: 4 metam-sodium applications).

also a function of a number of factors related to the characteristics of the building (27):

$$C_i \propto qkC_o/V$$

The factor 'q' is air flow rate (m^3/min), 'k' is a mixing factor to account for non-perfect mixing of air in a room, and 'V' is the volume of the room. The airflow rate 'q' is a function of the pressure drop 'Δp^n' across the building shell (30):

$$q \propto \Delta p^n$$

The exponent 'n' is a function of the type of flow (e.g., laminar, turbulent) through the building shell. Depending on the permeability of the building shell, the infiltration rate can fall in the range 0.07-0.39/hour (26). Discussions of infiltration assume that all doors and windows remain shut. Of course, two significant infiltration routes would be through doorways as occupants enter and exit buildings and through windows opened for ventilation.

Building infiltration was clearly shown by the correlations of indoor with outdoor concentrations of MITC measured during summer and winter months in a metam-sodium application region near Bakersfield, CA (Figure 3) (25). While there are food sources for MITC (horse radish, cruciform vegetables [31,32]), the strong correlations in Figure 3 indicate that indoor MITC was essentially due to infiltration by outdoor residues. We are aware of only one other study of an agriculturally applied pesticide that measured indoor air concentrations (26). Application of *Bacillus thuringiensis* var. *kurstaki* (*Btk*) for gypsy moth eradication resulted in measurable residues indoors, as shown by the indoor/outdoor ratios: 0.22 during application and 3.2, 5-6 hours after application. In this case, the investigators assumed that much of the relatively high indoor residue post-application was due to *Btk* residues carried indoors on the clothing of the investigators. One additional study that also showed infiltration by outdoor residues was for methyl bromide used as a structure fumigant (33). Again using indoor/outdoor ratios, a house near a tarped and fumigated house (15-30 meters) showed ratios in the range 0.75-1.5. There is obviously a serious lack of data regarding the occurrence and persistence of fumigants – and pesticides in general – in indoor air due to outdoor infiltration. This is an area of research that needs to be addressed and the lack of data redressed.

The results of the infiltration studies strongly suggest that staying indoors during soil fumigation may not afford any protection against exposure. Furthermore, the indoor/outdoor ratios greater than unity indicate that indoor residues tend to persist relative to outdoor residues. This is a tentative statement, since only one fumigant – MITC – has been simultaneously monitored in indoor and outdoor air environments during soil treatment. Outdoor airborne fumigants can be diluted by fresh air (indoor air tends to be stagnant) and they can also undergo chemical and photochemical conversions that would not be available to indoor residues. During the summer, for example, outdoor surfaces (soil, pavement) would be more reactive toward fumigants due to activation by heat

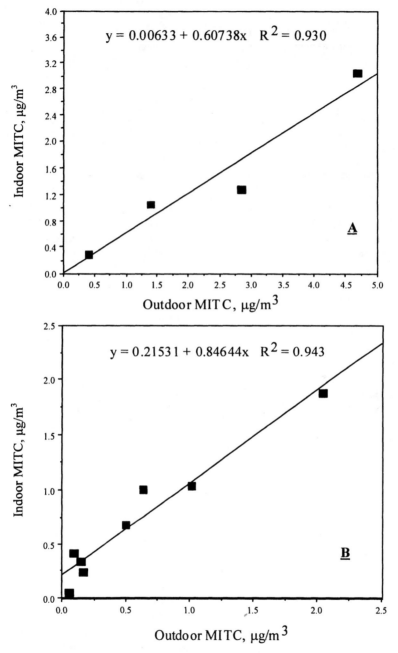

Figure 3. Relationship between indoor and outdoor MITC in air for
May-August (A) and January/March (B).

and sunlight. While indoor environments may contain some adsorptive and reactive surfaces, the relatively stagnant air conditions and less energetic artificial lighting would lead to slower dissipative processes. Of course, if external winds shift from fumigant sources, infiltration by relatively cleaner air would also eventually dissipate indoor residues.

Another correlation that has predictive potential with regard to exposure and risk assessment is illustrated in Figure 4, which shows correlations between total mass of metam-sodium applied and measured concentrations of MITC in indoor and outdoor air. This approach is similar to that of other investigators who made linear correlations of application data for methyl bromide with measured outdoor air concentrations to derive a predictive tool for exposure (*34*). However, we used Ln-Ln plots in Figure 4, since metam-sodium applications spanned several orders of magnitude. Once a reliable correlation has been established and validated using several seasons of data, all that is needed to assess outdoor exposure is to search application records. Figure 4 shows that this can also be done for indoor exposures as well.

Assessing Exposure

This brings us to the concept of hazard quotient (HQ), as defined by Lee et al. (*2*):

$$HQ = (measured\ exposure)/RfC$$

While HQ≥1 is not necessarily a cause for alarm, because of uncertainties associated with this term, it should be noted nonetheless, especially if HQ exceeds unity by a wide margin on a frequent basis. Table V summarizes MITC data for two monitoring seasons (sub-chronic exposure, RfC = 3 $\mu g/m^3$). For seven of the indoor sampling periods, sub-chronic HQ fell in the range 1.0-4.9; for six of the outdoor sampling periods, sub-chronic HQ fell in the range 1.0-5.4. For all of the winter sampling periods, sub-chronic HQ was somewhat less than unity (average HQ = 0.25-0.26). Lee et al. (*2*) obtained similar results for MITC (sub-chronic HQ = 2.1-8.5) – derived from numerous studies by other investigators. Table VI summarizes acute, sub-chronic, and chronic HQ data for four common soil fumigants taken from (*2*). Data for carbon disulfide are lacking. However, in light of the chronic RfC for carbon disulfide of about 700 $\mu g/m^3$ (Table II) and an 8-hour time-weighted-average baseline of 55,100 $\mu g/m^3$ (*35*), exposure under acute and sub-chronic conditions should not be any cause for concern.

Methyl bromide usage is slated for phase-out in the U.S. during 2005, except for critical use exemptions. As usage of this fumigant as a pre-plant treatment for soil declines, usage of other fumigants and their combinations will increase. Many of the substitute fumigants are not as efficacious as methyl bromide, even though some have acute and sub-chronic inhalation RfC values less than those for methyl bromide (Table II). So, increased amounts of the

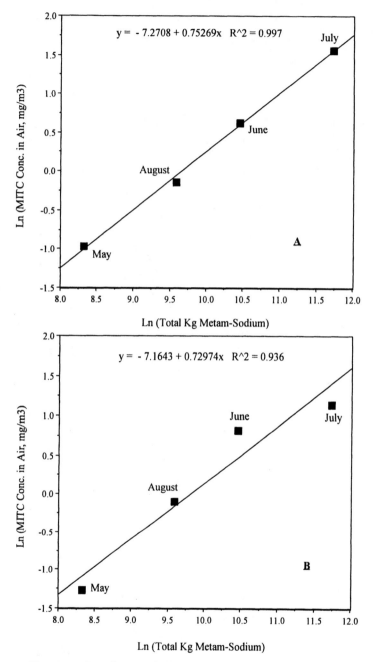

Figure 4. Correlation of MITC in outdoor (A) and indoor (B) air with metam-sodium application.

replacement fumigants may be needed to achieve the same pest control end point, leading to situations where the HQs may exceed unity on a regular basis. This takes on a special concern for children, who have consistently greater exposure risks compared to adults because of a greater inhalation to body weight ratio for children.

Table V. Sub-chronic hazard quotients (HQ)[a] for MITC.

Season	Category	Indoor[b]	Outdoor[c]
May-August (30 applications)	Air Concentration ($\mu g/m^3$)	0.08-14.7	0.06-16.3
	HQ range	0.03-4.9	0.02-5.4
January/March (4 applications)	Air Concentration ($\mu g/m^3$)	0.05-2.13	0.05-2.66
	HQ range	0.02-0.71	0.02-0.89

[a]HQ = (measured exposure)/RfC; RfC = 3 $\mu g/m^3$ (2).
[b]HQ ≥ 1 (1.0-4.9) for seven summer air concentrations.
HQ < 1 (average: 0.25) for all winter air concentrations.
[c]HQ ≥ 1 (1.0-5.4) for six summer air concentrations.
HQ < 1 (average: 0.26) for all winter air concentrations.

Table VI. Child/adult non-cancer hazard quotients (HQ) (2).

Fumigant	Acute	Sub-chronic	Chronic	Lifetime Cancer risk[a]
MITC	0.3-18.0	2.1-8.5	1.0-6.8	--
Methyl Bromide	0.005-0.7	0.007-13.9	0.003-2.0	--
1,3-D	0.002-0.5	0.02-11.5	0.001-2.0	8×10^{-7} to 3×10^{-4}
Chloropicrin	0.2	0.2-1.4	0.01-0.09	--

[a]95th percentile; risk = exposure x potency factor (= $4 \times 10^{-6}/[\mu g/m^3]$).

Remediations for Reduced Fumigant Emissions

Techniques have been developed for reducing emission losses of fumigants. The following is a summary of recommendations/suggestions:

1. Virtually impermeable film (VIF). One study (36) demonstrated that emission losses of methyl bromide could be reduced from 50-60%, using conventional 1-mil high-density polyethylene (HDPE), to less than 5% using VIF. Virtually impermeable film consists of a three-layer sandwich of low-density polyethylene for the outer layers and a center layer of gas-impermeable polyamide – permeability is 0.2 $g/m^2/hour$, which is about 300 times less than that for HDPE.

2. <u>Soil amendment with ammonium thiosulfate</u>. Halogenated fumigants (e.g., methyl bromide, 1,3-D, chloropicrin) can be readily dehalogenated by thiosulfate, as illustrated by the following reaction: $RX + S_2O_3^{2-} \rightarrow RS_2O_3^- + X^-$. In a field study where the soil surface was treated with ammonium thiosulfate solution (~660 kg/ha, a typical application rate for fertilizer) prior to methyl bromide fumigation by drip line, less than 10% of the applied fumigant was lost through emissions, compared to about 61% emission losses without amendment (*37*).

3. <u>'Capture and destroy'</u>. The idea is to divert halogenated fumigants from under the tarp to a chemical reactor or bioreactor. For example, one study demonstrated the facile dehalogenation of methyl bromide, propargyl bromide, 1,3-D, chloropicrin, and methyl iodide by thiosulfate in solution (*38*). For an 8:1 molar concentration ratio of thiosulfate to fumigant, half-lives ranged from about 1 hour (methyl bromide) to about 36 hours (chloropicrin). Private industry is developing systems for destroying methyl bromide that involve a chemical reactor that uses phase transfer catalysts (*39*) and a bioreactor consisting of biotrickling filters (*40*). The manufacturer of the chemical reactor claims a greater than 85% removal of methyl bromide from a forced air stream (fumigation chamber) in one pass through the reactor.

4. <u>Intermittent water seal for surface chemigation/shank injection</u>. This would have specific relevance to metam-sodium (MITC precursor) and sodium tetrathiocarbonate (carbon disulfide precursor) applications. A standard water seal involves three water applications starting at the end of the fumigant application. For MITC, water applications are spread out over a 30-hour period (0.5" per application; 1.5" total). With the standard water seal, MITC emission rates from both surface chemigation and shank injection are often greater than 300 $\mu g/m^2{\cdot}sec$. The proposed intermittent seal involves seven water applications, starting at the end of fumigant application, over about a 36-hour period (0.5" from the first watering followed by six 1/6" waterings; 1.5" total). The result is an order-of-magnitude reduction in the MITC emission rate: 23 $\mu g/m^2{\cdot}sec$ for shank injection and 93 $\mu g/m^2{\cdot}sec$ for surface chemigation (*41*).

For the halogenated fumigants, the most straightforward remediation approach would be the use of virtually impermeable films, since the application technology is already in place, with ammonium thiosulfate soil amendment as a close second. The 'capture and destroy' approach seems impractical under field conditions, while an intermittent water seal may be the only way to reduce MITC and carbon disulfide emissions. A significant reduction in fumigant emission losses would mean that less material would be needed to achieve the same level of pest control. This has obvious positive exposure and economic implications.

84

Conclusions

It is likely that, for the near future, fumigants will continue to be used to enhance agricultural productivity and quality. In so doing, some unintended exposure to humans, animals, and ecosystems will occur – this can't be completely avoided. The described measurement and modeling techniques for determining fumigant emissions and downwind concentrations have been or can be used for the assessment of inhalation exposure. Measurement techniques, for example, have shown that some current fumigant application practices lead to significant emission losses and to exposures – both outdoors and indoors (through building infiltration of outdoor residues) – that exceed acute and sub-chronic reference concentrations (RfCs) on a somewhat frequent basis. While this may not necessarily be a cause for alarm, the phase-out of methyl bromide – probably the best broad-spectrum fumigant in agriculture's arsenal – could lead to an increase in frequency of exposures exceeding RfCs because of the increased usage of other less efficacious fumigants, some of which have lower RfC values. It becomes imperative, then, that control measures, such as virtually impermeable film and soil surface treatments (chemical, water), be implemented at the source to minimize fumigant emission losses.

References

1. CDPR. California Department of Pesticide Regulation, private communication, 2005.
2. Lee, S., McLaughlin, R., Harnly, M., Gunier, R., and Kreutzer, R. *Environmental Health Perspectives*, 2002, *110*, 1175.
3. Majewski, M.S., McChesney, M.M., Woodrow, J.E., Pruger, J.H., and Seiber, J.N. *J. Environ. Qual.*, 1995, *24*, 742.
4. Yates, S.R., Gan, J., Ernst, F.F., Mutziger, A., and Yates, M.V. *J. Environ. Qual.*, 1996, *25*, 184.
5. Yates, S.R., Ernst, F.F., Gan, J., Gao, F., and Yates, M.V. *J. Environ. Qual.*, 1996, *25*, 192.
6. Seiber, J.N., Woodrow, J.E., Honaganahalli, P.S., LeNoir, J.S., and Dowling, K.C. In *Fumigants: Environmental Fate, Exposure, and Analysis*, J.N. Seiber, J.A. Knuteson, J.E. Woodrow, N.L. Wolfe, M.V. Yates, and S.R. Yates, 1997, 154.
7. Yagi, K., Williams, J., Wang, N.Y., and Cicerone, R.J. *Proc. Natl. Acad.*, 1993, *90*, 8420.
8. Yagi, K., Williams, J., Wang, N.Y., and Cicerone, R.J. *Science*, 1995, *267*, 1979.
9. Yates, S.R., Gan, J., Ernst, F.F., and Wang, D. *J. Environ. Qual.*, 1996.

10. Papiernik, S.K., Dungan, R.S., Zheng, W., Guo, M., Lesch, S.M., and Yates, S.R. *Environ. Sci. Technol.,* **2004**, *38*, 5489.

11. Ross, L.J., Johnson, B., Kim, K.D., and Hsu, J. *J. Environ. Qual.,* **1996**, *25*, 885.

12. EPA, 2005. U.S. Environmental Protection Agency, Support Center for Regulatory Air Models (SCRAM), http://www.epa.gov/ttn/scram/.

13. LE, 2005. Lakes Environmental, http://www.lakes-environmental.com.

14. Denmead, O.T., Simpson, J.R., and Freney, J.R. *Soil Sci. Soc. Am. J.,* **1977**, *41*, 1001.

15. Rolston, D.E. In *Methods of Soil Analysis, Part I. Physical and Mineralogical Methods,* American Society of Agronomy—Soil Science Society of America, **1986**.

16. Woodrow, J.E., Hebert, V., and LeNoir, J.S. In *Handbook of Residue Analytical Methods for Agrochemicals, Vol. 2,* P.W. Lee, H. Aizawa, A.C. Barefoot, and J.J. Murphy, **2003**, 908.

17. Biermann, H.W. In *Fumigants: Environmental Fate, Exposure, and Analysis,* J.N. Seiber, J.A. Knuteson, J.E. Woodrow, N.L. Wolfe, M.V. Yates, and S.R. Yates, **1997**, 202.

18. EPA, 1994. U.S. Environmental Protection Agency, *Emission Factor Documentation for AP-42, Section 9.2.2, Pesticide Application,* EPA contract 68-D2-0159, September.

19. Woodrow, J.E., Seiber, J.N., and Baker, L.W. *Environ. Sci. Technol.,* **1997**, *31*, 523.

20. Woodrow, J.E., Seiber, J.N., and Dary, C. *J. Agric. Food Chem.,* **2001**, *49*, 3841.

21. Sullivan, D.A., Holdsworth, M.T., and Hlinka, D.J. *Atmospheric Environment,* **2004**, *38*, 2471.

22. Reiss, R., and Griffin, J. *A Probabilistic Exposure and Risk Model for Fumigant Bystander Exposures Using Iodomethane as a Case Study,* Sciences International, Inc., **2004**.

23. CDPR. California Department of Pesticide Regulation, *Air Monitoring for Methyl Isothiocyanate During a Sprinkler Application of Metam-Sodium,* Report EH-94-02, **1994**.

24. Honaganahalli, P.S., and Seiber, J.N. *Atmospheric Environment,* **2000**, *34*, 3511.

25. Seiber, J.N., Woodrow, J.E., Krieger, R.I., and Dinoff, T. *Determination of Ambient MITC Residues in Indoor and Outdoor Air in Townships Near Fields Treated with Metam Sodium,* Final Report, June, **1999**.

26. Teschke, K., Chow, Y., Bartlett, K., Ross, A., and van Netten, C. *Environmental Health Perspectives,* **2001**, *109*, 47.

27. Tung, T.C.W., Chao, C.Y.H., and Burnett, J. *Atmospheric Environment,* **1999**, *33*, 881.

28. Lee, H.S., Kang, B.-W., Cheong, J.-P., and Lee, S.-K. *Atmospheric Environment,* **1997,** *31,* 1689.

29. Gibbons, D., Fong, H., Segawa, R., Powell, S., and Ross, J. *Methyl Bromide Concentrations in Air Near Fumigated Single-Family Houses,* Report HS-1713, California Department of Pesticide Regulation, **1996.**

30. Sirén, K. *Building and Environment,* **1993,** *28,* 255.

31. Fahey, J.W., Zalcmann, A.T., and Talalay, P. *Phytochemistry,* **2001,** *56,* 5.

32. Fenwick, G.R., Heaney, R.K., and Mullin, W.J. In *CRC Critical Reviews in Food Science and Nutrition,* **1983,** *18,* 123.

33. Li, L.Y., Johnson, B., and Segawa, R. *J. Environ. Qual.,* **2005,** *34,* 420.

34. EPA, 1989. U.S. Environmental Protection Agency, EPA/600/8-88/066F, August.

35. Wang, D., Yates, S.R., Ernst, F.F., Gan, J., and Jury, W.A. *Environ. Sci. Technol.,* **1997,** *31,* 3686.

36. Gan, J., Yates, S.R., Becker, J.O., and Wang, D. *Environ. Sci. Technol.,* **1998,** *32,* 2438.

37. Wang, Q., Gan, J., Papiernik, S.K., and Yates, S.R. *Environ. Sci. Technol.,* **2000,** *34,* 3717.

38. VR, 2005. Value Recovery, http://www.ptcvalue.com.

39. ERI, 2005. Energy Resource Institute, http://www.energy-institute.com.

40. Sullivan, D.A., Holdsworth, M.T., and Hlinka, D.J. *Atmospheric Environment,* **2004,** *38,* 2457.

Chapter 6

Setting Fumigant Application Buffer Zones

T. A. Barry, B. Johnson, and R. Segawa

California Environmental Protection Agency, Department of Pesticide
Regulation, Sacramento, CA 95812

California's Department of Pesticide Regulation (DPR) has
found unacceptable risk to human health associated with some
fumigant inhalation exposure. DPR uses air monitoring data
and computer modeling to estimate exposures. Monitoring
provides a snapshot of air concentrations in the vicinity of
specific pesticide applications. The Industrial Source
Complex-Short Term (ISCST) model, a Guassian Plume air
dispersion model, estimates air concentrations under a variety
of conditions. If monitoring data and computer modeling
indicate unacceptable air concentrations in the vicinity of
pesticide applications, DPR uses the ISCST model to
determine the appropriate size and duration of buffer zones.
These techniques are illustrated using methyl bromide as an
example.

Background

Fumigants are highly volatile pesticides applied to soil at high rates compared to other pesticides. Fumigants volatilize from soil and move off-site causing concern about potential health hazards associated with inhalation exposure. In 2003, four fumigants accounted for approximately 20% of the reported pounds of pesticides used in California: 1,3-Dichloropropene, Chloropicrin, Methyl Bromide, and Metam-Sodium and other Methyl Isothiocyanate (MITC) generating compounds. These fumigants are applied to soil by a variety of methods, including sprinkler, drip, and soil injection. A Methyl Bromide soil injection in California is shown in Figure 1.

Risk assessments evaluate potential health hazards associated with fumigant use. Toxicology data are evaluated and the No-Observed Adverse Effect Level (NOAEL) is determined. The toxicology data are typically from animal studies. Therefore, the NOAEL must be adjusted to account for differences between test animals and humans and variation in the human population. Through the risk assessment process, a human equivalent air concentration of the NOAEL is derived. The human equivalent air concentration is a Time Weighted Average (TWA) concentration consisting of exposure duration and concentration level that is appropriate based upon the toxicology of a particular fumigant. Exposure data (e.g., air concentrations) are then evaluated against the toxicology data and the human equivalent air concentration. Risk assessments have shown unacceptable exposures to some fumigants under some exposure scenarios. These unacceptable exposures require development of mitigation measures to reduce exposures.

Following completion of the risk assessment, the Department of Pesticide Regulation (DPR) initiates the risk management process. In the risk management process DPR management selects the Level of Concern (LOC) air concentration and acceptable probability of exceeding the LOC air concentration. For example, the DPR LOC air concentration for acute methyl bromide exposure is 210 ppb (815 $\mu g/m^3$) as a 24-hr TWA (1). Methyl Bromide air monitoring and air dispersion modeling were used to assess whether the LOC air concentration is likely to be exceeded and to develop and evaluate mitigation measures.

Air Monitoring

For an individual application, air samplers are placed from 10 m to 100 m from the edge of the field at 8 to 24 locations around the field. Sampling interval duration varies by fumigant according to the LOC air concentration averaging time. For example, the Methyl Bromide LOC averaging time of 24 hours while the MITC LOC averaging time is 8 hours (2). However, sampling intervals are typically 4 hrs to 24 hrs. The sampling is conducted for 2 to 14 days following the beginning of the application. Air monitoring is most often accomplished by

Figure 1. A methyl bromide shallow injection to a flat field followed by a tarp.

drawing air through sampling tubes filled with a trapping medium appropriate for the fumigant under study. For example, Methyl Bromide sampling uses petroleum based charcoal filled tubes while Chloropicrin sampling uses XAD-4 resin (macroreticular cross-linked aromatic polymer) filled tubes. Alternatively, air sampling can be conducted using stainless steel canisters. Meteorological data are also collected on-site. Figure 2 illustrates a sampler layout with measured air concentrations and a wind rose characterizing the wind conditions during a single sampling interval. A typical time trend in maximum concentration measured during each of the sampling intervals from a field study is shown in Figure 3. It should be noted that the maximum air concentration is not necessarily measured at the same sampler each interval. The sampler showing the maximum air concentration will depend primarily upon the predominant wind direction during the sampling interval.

DPR has data from over 40 methyl bromide fumigation air monitoring studies. Results from these studies demonstrate that measured air concentrations vary with many factors, including distance from the field, application method and rate, and meteorological conditions. Thus, a single monitoring study provides concentrations representing only that specific set of conditions. DPR supplements air monitoring data with air dispersion computer modeling to estimate air concentrations for other sets of conditions and to develop mitigation measures such as buffer zones.

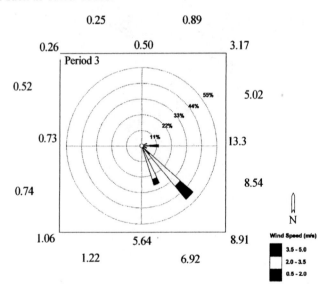

Figure 2. An example of key monitoring results for one sampling period. Air concentrations (ppb) measured at each air sampler are shown as well as a wind rose characterizing the wind conditions. The samplers are 15m and 45m from the field edge for the inner and outer rings, respectively.

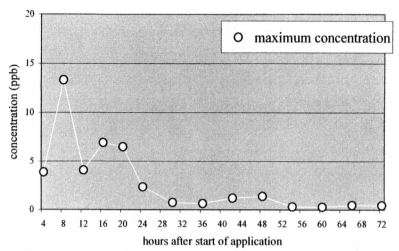

Figure 3. An example of key monitoring results for the duration of air monitoring. Shown is the maximum concentration measured during each sampling interval for 72 hours following the commencement of sampling.

Computer Modeling

DPR uses the U.S. Environmental Protection Agency Industrial Source Complex Short Term (ISCST3) model to develop and evaluate mitigation measures. The ISCST3 model is a Gaussian Plume model (3) that estimates air concentration (χ) for a ground level source at downwind distance x (m), crosswind distance y (m), and height z (m) (z = 0 at ground level) using the function shown below:

$$\chi(x,y,z) = \frac{Q}{2\pi\mu_s\sigma_y\sigma_z}\exp[-0.5\left(\frac{y}{\sigma_y}\right)^2]$$

Where:

χ = air concentration ($\mu g/m^3$)

x = distance downwind on the plume centerline (m)

y = distance cross-wind from the plume centerline (m)

z = distance from ground-level (m) (z = 0 at ground level)

Q = volatilization flux ($\mu g/m^2 s$)

σ_y = standard deviation of cross-wind concentration distribution at x, distance downwind (m)

σ_z = standard deviation of vertical concentration distribution at x, distance downwind (m)

μ_s = mean wind speed (m/s)

The terms σ_y and σ_z quantitatively characterize atmospheric mixing. See (4) for further information on the determination of the values for σ_y and σ_z. Typical adult breathing height is assumed to be $z = 1.2m$ to $z = 1.5m$. Air concentrations may be estimated for either ground level ($z = 0m$) or at a selected breathing height. For a ground level source, at a fixed downwind distance, maximum air concentrations are at ground level.

The volatilization flux is the most influential variable determining the air concentrations associated with a fumigant application. There are several methods to estimate the volatilization flux. Direct methods include the Aerodynamic-Gradient Technique (5), and the Integrated Horizontal Flux method (6). These methods measure air concentrations and wind speed at several heights directly above the source or at the downwind edge of the source and estimate the volatilization flux based upon the vertical profile of wind and air concentration. The direct methods require sensitive meteorological instrumentation and air concentration measurements. In addition, a minimum fetch (upwind length of the source) is required to obtain a reliable estimate of the volatilization flux. Chamber Methods (7) also measure volatilization flux directly by placing an open-bottom chamber over a small area of soil surface and measuring the gas emitted into the chamber. The chamber used may be a passive (closed) or an active (dynamic, flowing) sampling system. Chamber methods are much simpler than the gradient methods but do have drawbacks. The most significant drawback is that the presence of the chamber alters the relationship between the soil and the atmosphere.

An important property of the Gaussian Plume model is the proportional relationship between volatilization flux and air concentration. This relationship allows indirect estimation of the volatilization flux using the Back-calculation method (8). The Back-calculation method uses air concentration measurements from an air monitoring study designed as discussed above together with the ISCST3 model. A nominal volatilization flux value is used initially as input to run the ISCST3 model. The model estimates air concentrations at receptor locations corresponding to the location of air samplers during the air monitoring study. Regression analysis is then used to assess agreement between measured and modeled air concentrations. Provided wind speed conditions during the air monitoring sampling interval were not calm, the wind direction driven pattern of air concentrations estimated by the ISCST3 model will generally match that observed in the measured air concentrations. The match in magnitude between the measured and modeled air concentrations is achieved by multiplying the nominal volatilization flux by the slope of the regression line. The Back-calculation method is less expensive to conduct than direct methods and volatilization flux estimates obtained with this method compare favorably to the direct measurement aerodynamic method (9).

For regulatory purposes the volatilization flux is often expressed as an integrated mass fraction of the application rate (emission ratio) for the averaging time associated with the LOC air concentration. The methyl bromide emission ratio is the proportion volatilized during the peak 24-hour period because the LOC air concentration is expressed as a 24-hr TWA. DPR has used the Back-calculation method to estimate volatilization flux, and thus the emission ratio, for 43 methyl bromide applications (10). Figure 4 shows a box-plot summary of the methyl bromide emission ratios by application method. The four application methods shown in Figure 4 have statistically different mean emission ratios (11). These emission ratios were used to distinguish the application methods in Methyl Bromide buffer zones development and regulations.

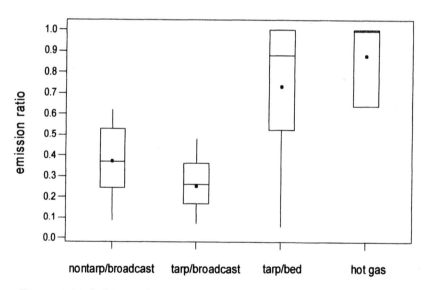

Figure 4. Methyl Bromide emission ratios grouped by application method according to statistically different mean emission ratios. The box deliniates the first (Q_1) to third quartile (Q_3), the horizontal line is the median, the dot is the arithmethic mean and the vertical lines show adjacent values within $\pm 1.5(Q_3 - Q_1)$.

94

Regulatory Requirements

DPR used the ISCST3 model to develop mitigation measures and to determine regulatory requirements for methyl bromide use. Volatilization flux estimates from the 43 monitoring studies together with the screening meteorological conditions of 1.4 m/s wind speed, constant wind direction, and C stability (slightly unstable atmospheric conditions) were used to find distances to the LOC air concentration. The screening meteorological conditions were chosen based upon DPR analysis of two initial studies where it was found that C stability and 1.4m/s adequately characterized 24-hr time weighted average air concentrations (12). Screening meteorological conditions were used because the regulations were to be applicable statewide. However, historical meteorological data (e.g., 5 years of data from a single meteorological station, or sets of data from multiple stations) could also be used. Output from these analyses consists of the required buffer zone distances necessary to maintain the failure rate of buffer zones below the risk management decision level (e.g. buffer zones long enough to capture the LOC air concentration in 95% of all applications in long term practice). Figure 5 illustrates the determination of the required buffer zone distance using output for one day (24-hr TWA) from ISCST3 modeling generated with historical meteorological data.

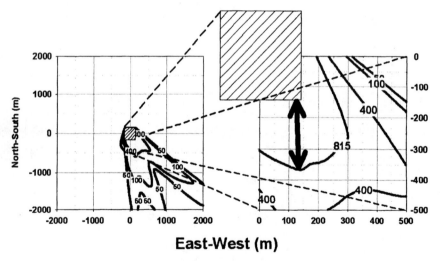

East-West (m)

Figure 5. ISCST3-modeled air concentration isopleths (μg/m³) obtained using 24 hours of hourly historical meteorological data. This figure illustrates the determination of the required buffer zone for a methyl bromide application using model generated air concentration isopleths. The field is 20 acres (hatched square). The largest distance (double ended arrow on right) between the field and the 815 μg/m³ isopleth is the required buffer zone.

Table I. Emission ratios of the three major Methyl Bromide application types allowed by California regulation.

Fumigation Method	Emission Ratio[1]
Shallow injection to flat field using "Noble plow" shank and tarp	0.25
Shallow injection to bedded field Using straight shank and tarp	0.80
Deep injection to flat field using Straight shank and no tarp	0.40

[1]maximum fraction volatilized in 24 hours

Table I lists the emission ratios for three methyl bromide application methods. These emission ratios are the mean values for the same three application methods shown in Figure 4. The first step in finding a required buffer zone is to specify the method of fumigation. Only methods that are similar to those monitored in the 43 air monitoring studies are allowed. The emission ratio for the application method is then used to look up the required buffer zone in the Methyl Bromide Field Fumigation Buffer Zone Determination, Est. 2/04 (13). Methyl Bromide buffer zone distances range from 50 to 4600 feet (15 to 1400 meters) depending upon the volatilization flux (application rate x emission ratio) and area treated. Buffer zone durations range from 36 to 84 hours. There are limits on the application rate and area treated. In addition, a separation of 1300 feet and 36 hours is required between applications.

Effectiveness of the methyl bromide buffer zones in mitigating off-site exposure was assessed by two methods. The first method tallied results from the air monitoring studies, comparing the regulatory buffer zone distance to the model-calculated buffer based on each study's particular on-site meteorology and receptor geometry (14). In 97% of the cases the regulatory buffer zone was longer than necessary to capture the LOC air concentration. A second assessment was performed using historical meteorological data and a range of volatilization flux values corresponding to the emission ratios specified in regulation (15). Twenty-five combinations of volatilization flux and acreage along with 5 years of historical meteorological data from each of 4 counties were used as input to the ISCST3 model. Daily (24-hr TWA) off-site air concentrations were obtained for each acreage/flux combination for a total of 20 years (7,300 days). The buffer zone required for the application on each day was found and used to compile a distribution of required buffer zones for each acreage/flux combination. The results showed that for between 89.2% and 100% of the days, depending upon the acreage/flux combination, the regulatory buffer zone developed using the screening meteorological conditions was longer than the buffer required by the historical meteorological data. Thus, the regulatory buffer zones developed using the screening level meteorological conditions were protective on average at approximately the 95% level.

Summary

DPR uses air monitoring and air dispersion computer modeling to determine if exposures exceed the LOC air concentration and to develop mitigation measures. Buffer zones are a major mitigation measure employed by DPR for mitigation of offsite fumigant concentration exposures. The LOC air concentration and volatilization flux are key factors in determining the need for and size of buffer zones. Buffer zone size, duration, and other requirements vary with LOC air concentration, volatilization flux, area treated, and method of application.

Literature Cited

1. Lim, L.O. California Environmental Protection Agency, Department of Pesticide Regulation. Sacramento, CA. Methyl Bromide Risk Characterization Document – Volume I - Inhalation Exposure. 2002. http://www.cdpr.ca.gov/docs/dprdocs/methbrom/rafnl/mebr_rcd.pdf
2. Gosselin, P.H. California Environmental Protection Agency, Department of Pesticide Regulation. Sacramento, CA. Risk Management Directive, Metam-sodium and other Methyl Isothiocyante-generating pesticides. 2002. http://www.cdpr.ca.gov/docs/empm/pubs/mitc/dirctv120202.pdf
3. U.S. Environmental Protection Agency, User's guide for the Industrial Source complex (ISC3) Dispersion Model. Volume II – Description of Model Algorithms. EPA-454/B-95-003b, 1995. http://www.epa.gov/scram001/userg/regmod/isc3v2.pdf
4. Turner, D. B. Workbook of Atmospheric Dispersion Estimates; Lewis Publishers; FL, 1994.
5. Majewski, M; Desjardins, R.; Rochette, P.; Pattey, E.; Seiber, J.; Glotfelty, D. Environ. Sci. Technol. 1993, 27, 121-128.
6. Wilson, J. D., Shum, W. K. N. Agric. Forest Meteor. 1992, 57, 281-295.
7. Reichman, R. ; Rolston, D.E. J. Environ. Qual 2002, 31, 1774-1781.
8. Johnson, B., Barry, T., Wofford, P. California Environmental Protection Agency, Department of Pesticide Regulation. Sacramento, CA. Workbook for Gaussian modeling analysis of air concentration measurements. EH99-03, 1999. http://www.cdpr.ca.gov/docs/empm/pubs/ehapreps/eh9903.pdf
9. Ross, L. J.; Johnson, B.; Kim, D. K.; Hsu, J. J. Environ. Qual. 1996, 25, 885-891.

10. Segawa. R.; Johnson, B.; Barry, T. California Environmental Protection Agency, Department of Pesticide Regulation. Sacramento, CA, Summary of off-site air monitoring for methyl bromide field fumigations, Memorandum to John Sanders, January 21, 2000.
http://www.cdpr.ca.gov/docs/empm/pubs/tribal/1offsiteMonitMebr-2.pdf

11. Barry, T.A. California Environmental Protection Agency, Department of Pesticide Regulation. Sacramento, CA. Methyl bromide emission ratio groupings. Memorandum to Randy Segawa, December 2, 1999.
http://www.cdpr.ca.gov/docs/empm/pubs/tribal/methbrom%20emiss%20groupings.pdf

12. Johnson, B.; Barry, T. California Environmental Protection Agency, Department of Pesticide Regulation. Sacramento, CA. Estimation of Methyl Bromide flux for Seimer studies TC199.1 and F1.1 and use of standard weather conditions. Memorandum to Randy Segawa, October, 2005.
http://www.cdpr.ca.gov/docs/empm/pubs/tribal/EST_TC199All.pdf

13. California Department of Pesticide Regulation. Methyl Bromide Soil Funigation Buffer Zone Determination, Feburuary, 2004.
http://www.cdpr.ca.gov/docs/legbills/03004buffer_zones.pdf

14. Segawa, R.; Barry, T.; Johnson, B. California Environmental Protection Agency, Department of Pesticide Regulation. Sacramento, CA. Recommendations for methyl bromide buffer zones for field fumigations, Memorandum to John Sanders, January 21, 2000.
http://www.cdpr.ca.gov/docs/empm/pubs/tribal/1offsiteMonitMebr-2.pdf

15. Johnson, B. California Environmental Protection Agency, Department of Pesticide Regulation. Sacramento, CA, Evaluating the effectiveness of methyl bromide soil buffer zones in maintaining acute exposures below a reference air concentration, EH 00-10, 2000.
http://www.cdpr.ca.gov/docs/empm/pubs/ehapreps/eh0010.pdf

Chapter 7

Developing Pesticide Exposure Mitigation Strategies

Thomas Thongsinthusak and Joseph P. Frank

Worker Health and Safety Branch, California Department of Pesticide
Regulation, Cal/EPA, Sacramento, CA 95812

As part of the regulatory process, the California Department of
Pesticide Regulation (CDPR) develops mitigation strategies
when risk assessments identify unacceptable pesticide
exposure levels. These strategies must not only reduce
exposures to acceptable levels, they must also be practical and
enforceable. Typical mitigation measures can include personal
protective equipment, engineering controls, buffer zones,
lengthened reentry times, or restrictions on activities or use.
This chapter presents general considerations in developing
mitigation strategies for handlers (application related
exposures) and reentry workers (postapplication exposures),
and then provides mitigation measures for S-ethyl
dipropylthiocarbamate (EPTC) and methyl bromide. While
exposure issues were addressed with label changes for EPTC,
the strategies necessary to address methyl bromide exposures
were far more complex. They included a multistage process
that started with use-specific permit conditions, followed by
the development of new regulations intended to address field
fumigation applications. CDPR's mitigation strategies involve
intra- and inter-departmental, as well as public participation.

Exposure mitigation, which is part of the California Department of Pesticide Regulation (CDPR) risk management process, is undertaken in response to unacceptable risks identified in the risk assessment process. The primary objective of the mitigation strategy is to reduce exposures (dose) in order to reduce potential adverse responses (toxicologic effects). The typical mitigation strategy can include engineering controls, protective equipment, or use restrictions. These exposure reduction measures can be prescribed through label changes, permit conditions, or regulations. The target population for the mitigation strategy includes all individuals who may come into contact with pesticide products as a result of their registered use.

This chapter presents typical measures implemented to mitigate dermal and inhalation exposures that result from the use of pesticides in production agriculture. Mitigation strategies utilized for S-ethyl-dipropylthiocarbamate (EPTC) and methyl bromide are presented as examples. EPTC is used as a pre- and postplant herbicide to control annual grass, broadleaf weeds, and perennials. From 2001 to 2003, the average annual use of EPTC in California was 224,142 pounds of active ingredient (1). Within the same period, there was only one reported illness/injury (2). Methyl bromide is widely used as a fumigant to control pests in soil, fresh and dry agricultural products, residences, and other structures. From 2001 to 2003, the average annual use of methyl bromide in California was 6,864,919 pounds of active ingredient (1). Within the same period, there were four reported illnesses/injuries (2). For both active ingredients, the individuals considered to have the highest exposure potential were handlers (individuals involved with the application of the pesticide) and reentry workers. In the following sections, general and specific mitigation measures for pesticides are presented.

Selection Criteria for Consideration in Developing Mitigation Measures

The CDPR considers the following selection criteria when developing mitigation measures: (1) Are the measures effective and efficient in reducing exposures in order to reduce risks? (2) Are personal protective equipment (PPE) and engineering controls readily available? (3) Will the measures cause minimum or no health effects (heat stress, very uncomfortable)? (4) Are the measures economically feasible? (5) Is compliance practical? (6) Are the requirements enforceable? With regard to enforcement, the CDPR must consider whether CDPR staff or officials from California county agricultural commissioners will be able to monitor for compliance.

Potential Exposure Mitigation Measures for Handlers

One or more mitigation measures may be required to reduce exposure to an acceptable level when consider the above six selection criteria. General mitigation measures for handlers are the following: (1) PPE, e.g., coveralls, respirators chemical-resistant suits, gloves, footwear, aprons, and headgear; protective eyewear). (2) Engineering controls, e.g., closed mixing/loading system, enclosed cab with positive pressure and a charcoal air-filtration unit, water-soluble packaging. (3) Limit exposure time or amount of active ingredient handled. (4) Establish buffer zones.

Typically, the simplest approach is to use PPE. This is not, however, always practical. For example, while a full-body chemical-resistant suit can provide significant exposure reduction, it can also cause heat stress. In California, full-body chemical-resistant suits can only used when the temperature is at or below 80 °F during daylight hours (85 °F at night). Exposure reduction can also be accomplished by the use of closed mixing or pouring systems. A closed system is required in California when mixing liquid pesticides with the signal word "DANGER" on the label (toxicity category I) or when mixing Minimal Exposure Pesticides. Water-soluble packages effectively reduce dust for dry formulations of pesticides. A buffer zone (restricted area surrounding a pesticide application block) is particularly effective in reducing inhalation exposure to volatile pesticide such as fumigants.

Potential Exposure Mitigation Measures for Field-Workers

One or more mitigation measures may be required to reduce exposure of field-workers to pesticides to an acceptable level. The following are typical mitigation measures for field-workers that incorporate the above six selection criteria: (1) Establish a restricted entry interval (REI). (2) Require the use of engineering controls such as mechanical harvesters or pruners. (3) Limit the number of crops being treated. In addition to having an impact on dermal exposure for field-workers, this mitigation measure typically reduces dietary exposure to the public. (4) Require the use of PPE. It should be noted that reentry exposure concerns are often mitigated by REI and engineering controls.

The establishment of a safe REI, which is the time period after the application of pesticides when reentering the treated field is restricted, may be the most effective mitigation measure for reentry workers. This time period allows dislodgeable foliar residues (DFR) to dissipate or attenuate to an acceptable exposure level. With some crops, certain low contact activities are essential during the REI. An example would be some irrigation activities. When exceptions are allowed for entry into treated areas prior to the expiration of the REI, specific PPE requirements such as gloves and coveralls are used to minimize potential exposure. Another example where additional PPE is used to protect early entry workers is with growers that may have to harvest flowers on a daily basis, even though pesticides are used. For some crops, mechanical harvesters can be used to reduce or eliminate potential exposure. Examples include mechanical harvesting of tomatoes for processing/canning, harvesting certain varieties of nuts, and pruning various fruit trees. These cultural practices can greatly reduce dermal and inhalation exposures of field-workers.

When developing mitigation measures, it is important to factor in the level of protection provided by different types of work clothing, PPE, and engineering controls. When specific penetration values are not known, the CDPR uses the default protection factors shown in Table I (3). For convenience, default percent protection is used to determine the level of exposure reduction. Depending on the PPE used, exposure levels can be reduced by as much as 98%. When chemical-specific protection values are available, they are typically used rather than the default percent protection.

If a default protection factor is not known for specific PPE, application equipment, or modified application equipment, a protection factor can be experimentally derived. Alternately, actual field exposure data collected during the use of modified equipment may be employed to determine if the exposure levels are acceptable. For example, with methyl bromide, the CDPR has concluded that recent application tractor and attachment modifications have in fact reduced exposure potential. Therefore, measured air concentrations from studies using the modified application equipment were used to improve the exposure estimates. When application modifications produce an exposure estimate that is still unacceptable, other mitigation measures are considered.

Table I. Default Percent Protection

Engineering Controls, Work Clothing, and PPE[a]	Protection (%)
Engineering Controls	
Closed mixing/loading system	95
Enclosed cab with positive pressure and a charcoal air-filtration unit	98
Enclosed cab	90
Work Clothing and PPE	
Work clothing such as long-sleeved shirt, long pants	90
Coveralls or overalls	90
Chemical-resistant full-body suit	95
Chemical-resistant gloves	90
Full-face respirator with cartridges approved by NIOSH/MSHA	98
Half-face respirator with cartridges approved by NIOSH/MSHA	90

[a] National Institute for Occupational Safety and Health (NIOSH). Mine Safety and Health Administration (MSHA). Engineering controls protect dermal and inhalation exposures; whereas, work clothing and PPE protect either dermal or inhalation exposure based on the intended use.

Risk Assessment and Mitigation Measures for EPTC

The risk assessment for EPTC was performed for acute, seasonal (subchronic), and annual (chronic) exposures (4) using estimated exposures for handlers (5). Table II shows the critical toxicity endpoints for EPTC that were used in the CDPR's risk assessment. The highest experimental no-observed-effect level (NOEL) for acute exposure was 20 mg/kg body weight (BW)/day; whereas, the lowest NOEL for annual exposure was 500 μg/kg BW/day. These experimentally determined NOELs were used to calculate margins of exposure (MOE) for different work tasks and exposure scenarios by dividing the NOEL by the exposure estimate (MOE = NOEL/Exposure estimate).

Table II. Toxicity of EPTC Used in the Risk Assessment

Scenario	Experimental NOEL[a]	Effects in Animal Studies
Acute exposure	20 mg/kg BW/day	Neurotoxicity in rats
Seasonal exposure	700 μg/kg BW/day	Nasal cavity degeneration/ Hyperplasia in rats
Annual exposure	500 μg/kg BW/day	Neuromuscular degeneration in rats

[a] Toxicity studies obtained from pesticide registrants were summarized in (4).

Exposure estimates were based on 2 to 8 hours per workday, 6 to 8 working days in a 17-day season, and 6 to 16 working days per year depending on exposure scenarios. After consideration of potential toxicity issues and seasonal exposure potential for EPTC, the 17-day season was adjusted to 21 days during the mitigation process.

Table III shows a summary of exposure ranges and MOEs for EPTC. Typically, an exposure estimate that translates to an MOE of 100 or more is considered acceptable when the NOEL is based on animal data. The results of the EPTC risk assessment indicates that several seasonal exposure scenarios have MOEs lower than 100. Thus, exposure mitigation measures are needed.

Development of Mitigation Measures and Implementation for EPTC

EPTC mitigation measures were developed for dermal and inhalation exposures (6). These mitigation measures are in addition to what appear on federal product labels. They include long-sleeved shirt, long pants, chemical-resistant gloves, chemical-resistant apron, chemical-resistant footwear, socks, and protective eyewear. In addition to the PPE, the mitigation measures proposed limited daily work hours (6). During the risk management comment

Table III. Summary of EPTC Exposure Ranges and Margins of Exposure[a]

Work Task	Absorbed Dosage (µg/kg BW/day)		
	ADD (MOE)	SADD (MOE)	AADD (MOE)
1. Liquid Formulation:			
Ground Application			
Mixer/Loader	48.0 (417)	23.2 (30)	1.53 (328)
Applicator	21.3 (940)	10.7 (66)	0.94 (532)
Mixer/Loader/Applicator	90.9 (220)	43.4 (16)	4.43 (113)
2. Liquid Formulation:			
Water-Run			
Applicators	6.54 (3,059)	3.40 (206)	0.60 (830)
3. Liquid Formulation:			
Center-Pivot Sprinkler System			
Mixer/Loader/Applicator	222 (90)	79.2 (9)	4.13 (121)
4. Granular Formulation:			
Flowers/Ornamentals			
Loader/Applicator	15.5 (1,288)	7.94 (88)	0.81 (614)
5. Granular Formulation:			
Aerial Application			
Pilots	2.87 (6,974)	1.98 (353)	0.54 (932)
Flaggers	9.60 (2,084)	5.15 (136)	0.68 (731)
Loaders	86.4 (232)	41.3 (17)	2.37 (211)

Note: Absorbed daily dosage (ADD) = (Daily exposure x absorption rate)/BW; Seasonal average daily dosage (SADD) = ADD x Workdays in a season/21 days; Annual Average Daily dosage (AADD) = ADD x Workdays in a year/365 days.
[a] Exposures include dermal, inhalation, and dietary exposures. Absorbed dosages were calculated based on 18.25% dermal absorption and 50% respiration uptake (5).

period, a combination of PPE and the amount of EPTC handled by a handler per day and per season was suggested as an alternative to the PPE and limited daily work hours restrictions. The amount handled was considered approximately equivalent to the amount that would be handled in the proposed daily work hours. Table IV shows the adopted mitigation measures for EPTC. These measures were incorporated into product labels as additional requirements for California.

Risk Assessment and Mitigation Measures for Methyl Bromide

Inhalation exposures of handlers and residents/bystanders to methyl bromide were estimated for acute (daily), subacute (7 days), subchronic (90 days), and chronic (365 days) (7). Exposure data were obtained from field studies using various application methods. The 95th percentiles values were used for acute exposure estimates; whereas, the arithmetic means were used for subacute, subchronic, and chronic exposure estimates. Table V shows seasonal (subchronic) and acute exposure estimates of handlers for various work tasks and

Table IV. Mitigation Measures for EPTC in California[a]

Work Tasks/Formulations	Mitigation Measures
Handlers	Coveralls plus a half-face respirator[b]
Handlers (center pivot or CP)	Chemical-resistant full-body protective clothing and a half-face respirator[b]
Liquids:	
Mixer/loader	75 gal/day or 500 gal/21-day period
Applicator uses CP	20 gal/day or 40 gal/21-day period
Applicator in an enclosed cab	40 gal/day or 280 gal/21-day period
Applicator uses other application equipment	30 gal/day or 210 gal/21-day period
Granules:	
Handlers	100 lb/day or 1,000 lb/21-day period

[a] Product labels indicate "the operator of the property shall include in their pesticide use records the name(s) of person that handled the product for each application."

[b] Approved respirator with: an organic-vapor (OV) removing cartridge with a prefilter approved for pesticides (MSHA/NIOSH approval number prefix TC-23C) or a canister approved for pesticides (MSHA/NIOSH approval number prefix TC-14G) or a NIOSH approved respirator with an OV cartridge; or a canister with any N, R, P, or HE prefilter.

Table V. Ranges of Acute and Seasonal Exposures[a]

Application Methods	Seasonal (Average)	Acute (95th Percentile)
1. Nontarpaulin/Shallow/Bed:		
Applicator	416	828
Cultipacker driver	327	866
2. Nontarpaulin/Deep/Broadcast:		
Applicator	382	1,680
Cultipacker driver	202	1,084
Supervisor	280	741
3. Tarpaulin/Shallow/Broadcast:		
Applicator	459	1,124
Copilot	926	1,961
Shoveler	609	1,525
4. Tarpaulin/Shallow/Bed:		
Applicator	203	873
Copilot	503	1,114
Tarp cutter	326	1,210
Tarp remover	131	734
5. Tarpaulin/Deep/Broadcast:		
Applicator	459	1,124
Copilot	926	1,961
Shoveler	609	1,525
6. Drip System – Hot Gas:		
Applicator	793	2149

[a] Represents non-Time-Weighted-Average (non-TWA) air concentrations (parts per billion, ppb).

application methods. Other exposure estimates are not shown because mitigation measures have been developed only for acute and seasonal exposures.

The risk assessment for methyl bromide was performed for inhalation exposures (*8, 9*). The experimental NOELs and observed effects in animal studies are shown in Table VI. The reference concentration (RfC) was calculated from the human equivalent NOEL divided by an uncertainty factor (UF) of 100. The human equivalent NOEL was determined by incorporating breathing rates and body weights of humans and animals (*8, 9*). Results of the risk assessment indicated that MOEs for several exposure scenarios are lower than an acceptable level of 100. Therefore, mitigation measures were considered necessary.

Table VI. Risk Assessment for Methyl Bromide

Scenarios	Experimental NOEL	Effects in Animal Studies
Acute	40 ppm	Developmental toxicity (pregnant rabbits) RfC^a =210 ppb
Subchronic (6 weeks)	5 ppm	Neurotoxicity (dogs) RfC^b = 16 ppb (adult), 9 ppb (child)
Chronic	0.3 ppm (ENOEL)	Nasal epithelial hyperplasia/degeneration (rats) RfC^c =2 ppb (adult'). 1 ppb (child)

Note: Reference concentration (RfC) = Human equivalent NOEL/UF 100 (shown as the 24-h TWA).

[a] 24-hour TWA. Human equivalent NOEL/100 or 21 parts per million (ppm)/100 = 210 ppb (*8, 9*).

[b] 24-hour TWA. Human equivalent NOEL/l00 or 1.56 ppm/100 = 16 ppb (adult) and 0.88 ppm/100 = 9 ppb (child) (*9*).

[c] 24-hour TWA. Human equivalent NOEL/100 or 0.2 ppm/100 = 2 ppb (adult) and 0.1 ppm/100 = 1 ppb (child) (*8, 9*). ENOEL is estimated NOEL.

Development of Mitigation Measures and Regulations for Methyl Bromide

In 1993, the CDPR issued methyl bromide proposed soil injection fumigation permit conditions in order to reduce exposures of handlers to methyl bromide during field fumigation. From 1994 to 1997, the permit conditions were revised several times (*10*). The permit conditions included restrictions on types of application equipment, daily work hours, acreage treated, and buffer zones. In 2000, mitigation measures for acute inhalation exposures were developed (*11*) based on the results of the risk assessment. The mitigation measures included restrictions on types of application equipment, acreage treated, buffer zones, and daily work hours for handlers, which ranged from 2 to 7 hours. In the same year, the CDPR re-adopted emergency regulations, which became effective on January 21, 2003.

In 2003, mitigation measures for seasonal exposures were developed (*12*) based upon the RfC of 16 ppb (adult) (*9*). These mitigation measures included recommendations for application equipment, PPE, daily work time, buffer zones, tarpaulin cuffing method, and aeration time. Thereafter, the CDPR proposed permanent regulations, which included mitigation measures for acute and seasonal exposures during field fumigations. In 2004, the CDPR issued the permanent regulations for field soil fumigation (*13*) after the Office of Administrative Law approved the regulations.

The following are definitions of exposure periods used in the development of mitigation measures and calculations of work hours for handlers. A short-term (acute) exposure for methyl bromide was defined as the exposure that occurs from 1 to 7 days; whereas, an intermediate-term (seasonal) exposure was assumed to occur in a 30-day period. The seasonal exposure was assumed to be continuous and was compared to subchronic toxicity data. Since the potential exposure issues and necessary mitigation measures were different for acute and seasonal exposure, the CDPR attempted to identify situations where seasonal exposure potential would be considered negligible. After reviewing the toxicology profile and exposure potential for methyl bromide, the CDPR concluded that if a worker handled methyl bromide no more than three workdays in a calendar month, that any potential hazard was related to acute toxicity. Therefore, the risk potential for these workers could be mitigated with those measures put in place for acute exposure. When workers were exposed to more than 3 workdays in a calendar month, mitigation measures intended to control seasonal exposures must also be employed.

Calculations of maximum work hours and adjustment of work hours allowed by the regulations are as follows:

Determination of work hours for acute exposures: The maximum work hours (h/day) = [(210 ppb x 24 h)/95th Percentile breathing zone methyl bromide concentrations (ppb, not the TWA)].

Determination of work hours for seasonal exposure: The maximum work hours (h/day) = [(16 ppb x 24 h)/Arithmetic mean breathing zone methyl bromide concentrations (ppb, not the TWA)].

Revision of maximum work hours permitted by the regulations: [(Maximum application rate for method x Maximum work hours in a 24-hour period)/Actual application rate].

Maximum work hours for acute exposure based on different application methods and work tasks are shown in Table VII. Handlers can work from 2 to 7 hours per workday and a respirator is not required. For mitigation of seasonal exposures, handlers can handle methyl bromide according to the specified work hours (Table VIII). The handlers must wear NIOSH-certified respiratory protection specifically recommended by the manufacturer for use in atmospheres containing less than five ppm methyl bromide.

The regulations (*13*) describe other requirements in detail. These requirements include specific field fumigation methods, tarpaulin cutting and removal, tarpaulin repair, field fumigation notification, and buffer zones.

Table VII. Maximum Work Hours in a Maximum Three Workdays
Per Calendar Month (Acute Exposure)

Application Methods[a] and Work Tasks	Maximum Active Ingredient/Acre	Daily Maximum Work Hours[b]
1. Nontarpaulin/Shallow/Bed:		
Tractor Equipment Driving	200 lb	4[c]
Supervising		4[c]
2. NontarpaulinlDeep/Broadcast:		
Tractor Equipment Driving	400 lb	4[c]
Supervising		7[c]
3. Tarpaulin/Shallow/Broadcast:		
Tractor Equipment Driving		4[c]
Shoveling, Copiloting	400 lb	3[c]
Supervising		3[c]
Tarpaulin Cutting		4[d]
Tarpaulin Removal		7[d]
4. Tarpaulin/Shallow/Bed:		
Tractor Equipment Driving		4[c]
Shoveling, Copiloting	250 lb	4[c]
Supervising		4[c]
5. Tarpaulin/Deep/Broadcast:		
Tractor Equipment Driving		4[c]
Shoveling, Copiloting	400 lb	3[c]
Supervising		3[c]
6. Drip System – Hot Gas:		
Applicators		2[c]
Supervising	225 lb	2[c]

[a] Specific requirements are described in the regulations (13).

[b] Handlers are not required to wear a respirator.

[c] Work hours can be adjusted by using the formula shown above.

[d] Same work hours for these two work tasks are used for methods 4, 5, and 6.

Table VIII. Maximum Work Hours in a 24-Hour Period
(Seasonal Exposure)

Application Methods[a] and Work Tasks	Maximum Active Ingredient/Acre	Maximum Work Hours in a 24-h Period[b]
1. NontarpaulinlShallow/Bed:		
Tractor Equipment Driving	200 lb	9[c]
Supervising		9[c]
2. Nontarpaulin/Deep/Broadcast:		
Tractor Equipment Driving	400 lb	10[c]
Supervising		No limitation[d]
3. Tarpaulin/Shallow/Broadcast:	8C	
Tractor Equipment Driving		8[c]
Shoveling, Copiloting	400 lb	4[c]
Supervising		4[c]
Tarpaulin cutting		no limitation[d]
Tarpaulin removal		no limitation[e]
4. Tarpaulin/Shallow/Bed:		
Tractor Equipment Driving	250 lb	no limitation
Shoveling, Copiloting		8[c]
Supervising		8[c]
5. Tarpaulin/Deep/Broadcast:		
Tractor Equipment Driving		8[c]
Shoveling, Copiloting	400 lb	4[c]
Supervising		4[c]
6. Drip System – Hot Gas:		
Applicators		5[c]
Supervising	225 lb	5[c]

[a] Specific requirements are described in the regulations (13).

[b] Handlers are required to wear a half-face respirator.

[c] Work hours can be adjusted by using the formula shown above.

[d] Exception: An employee may perform this activity without a half-face respirator provided the employee does not work more than one hour in a 24-hour period. The maximum one-hour work limitation may be increased in accordance with the formula shown above. The same work hours are applied for tarpaulin cutting for application methods 4, 5, and 6.

[e] Exception: An employee may perform this activity without a half-face respirator provided the employee does not work more than three hours in a 24-hour period. The maximum three-hour work limitation may be increased in accordance with the formula shown above. The same work hours are applied for tarpaulin removal for application methods 4, 5, and 6.

Conclusions

Exposure mitigation measures for handlers and field-workers can be simple or complex depending on several factors such as the magnitude of unacceptable MOEs (i.e., very low MOEs require dramatic exposure reduction strategies that may not be practical), availability and affordability of PPE and engineering controls, availability of alternatives, economic importance of a pesticide. If reasonable mitigation measures cannot be developed, the use of that pesticide may need to be phased out or cancelled.

The CDPR adopted practical mitigation measures for EPTC and methyl bromide as demonstrated above. Implementation of mitigation measures for EPTC was accomplished by including those requirements on product labels. Development of mitigation measures for methyl bromide was more complex and involved high level administrators, staff of several branches, California agricultural commissioners, industry, lawyers, other agencies, and the public. Mitigation measures were initially put into use in permit conditions, emergency regulations, and finally in permanent regulations.

Acknowledgements

The authors would like to thank past and present staff of the California Department of Pesticide Regulation for their contributions, particularly Earl Meierhenry, the author of the EPTC risk characterization document, Lori Lim, the author of the methyl bromide risk characterization document, Randall Segawa and Charles M. Andrews for their helpful recommendations during the mitigation and regulatory process, Denis Gibbons, Robert I. Krieger and John H. Ross for their recommendations during the early stage of exposure assessment and mitigation process. The authors would also like to acknowledge contributions from numerous researchers and registrants who conducted exposure and toxicological studies. Authors of those studies were referenced in exposure and risk characterization documents for EPTC and methyl bromide.

References

1. Summary of Pesticide Use Report Data, Indexed by Chemical; California Department of Pesticide Regulation: Sacramento, CA, 2002-005.
2. Pesticide Illness Surveillance Program; California Department of Pesticide Regulation: Sacramento, CA, 2001-2003.
3. Thongsinthusak, T.; Ross, J. H.; Meinders, D. "Guidance for the Preparation of Human Pesticide Exposure Assessment Document"; Technical Report No. HS-1612; California Department of Pesticide Regulation: Sacramento, CA, 1993.

110

4. Meierhenry, E. F. "EPTC (S-Ethyl-Dipropylthiocarbamate) Risk Characterization Document"; California Department of Pesticide Regulation: Sacramento, CA, 1995.

5. Brodberg, R. K.; Thongsinthusak, T. "Estimation of Exposure of Persons in California to Pesticide Product Containing EPTC"; Technical Report No. HS-1531; California Department of Pesticide Regulation: Sacramento, CA, 1995.

6. Thongsinthusak, T. "Revised EPTC Mitigation Document"; Technical Report No. HSM-98002; California Department of Pesticide Regulation: Sacramento, CA, 1998.

7. Thongsinthusak, T.; Haskell, D. "Estimation of Exposure of Persons to Methyl Bromide During and/or After Agricultural and Nonagricultural Uses"; Technical Report No. HS-1659; California Department of Pesticide Regulation: Sacramento, CA, 2002.

8. Lim, L. O. "Methyl Bromide Risk Characterization Document: Inhalation Exposure (Volume I)"; Technical Report No. RCD 2002-03; California Department of Pesticide Regulation: Sacramento, CA, 2002.

9. Lim, L. O. "Methyl Bromide Risk Characterization Document: Inhalation Exposure (Addendum to Volume I)"; California Department of Pesticide Regulation: Sacramento, CA, 2003.

10. "Methyl Bromide Proposed Soil Injection Fumigation Permit Conditions"; California Department of Pesticide Regulation: Sacramento, CA, 1993 to 1997.

11. Gibbons, D.; Thongsinthusak, T. "Recommended Worker Safety Mitigation Measures for Methyl Bromide Soil fumigation Regulations"; Technical Report No. HSM-00014; California Department of Pesticide Regulation: Sacramento, CA, 2000.

12. Thongsinthusak, T.; Frank, J. P. "Mitigation Measures for Seasonal Exposures of Agricultural Workers to Methyl Bromide During Soil Fumigations"; California Department of Pesticide Regulation: Sacramento, CA, 2003.

13. "California Codes of Regulations (Title 3. Food and Agriculture), Division 6. Pesticides and Pest control Operations: Chloropicrin and Methyl Bromide-Field Fumigation"; California Department of Pesticide Regulation: Sacramento, CA, 2005.

Chapter 8

A Reality Fix for Risk Managers

John H. Ross[1], JJeffrey H. Driver[2], and Robert I. Krieger[3]

[1]infoscientific.com, Inc., 5233 Marimoore Way, Carmichael, CA 95608
[2]infoscientific.com, Inc., 10009 Wisakon Trail, Manassas, VA 20111
[3]Personal Chemical Exposure Program, Department of Entomology, University of California, Riverside, CA 92521

Human health risk assessment is the process of comparing hazard to exposure. In its quantitative form, human health risk analyses represent the systematic evaluation of the likelihood of an adverse effect arising from exposure within a defined population. Predictive human exposure and health risk analyses include uncertainties and often widely differing degrees of conservative bias. In the absence of the benefit of transparent and quantitative disclosure of variability and uncertainty, risk managers may be faced with making decisions on the basis of risk assessments that are significantly biased by the "Precautionary Principle,." with the philosophy that it is better not to allow the proposed use of a chemical than risk uncertain but possibly very negative consequences. In order to understand the uncertainties and conservative biases, it is useful to compare alternative risk management policies. For example, potential health risks associated with the use of common pharmaceuticals versus those associated with environmental exposures to pesticides indicates that more stringent "acceptable risk standards" are applied to pesticides. To illustrate the impact of conservative biases often applied to pesticides, this chapter summarizes comparisons between predictive models and actual measurements. Numerous variables used in "screening-level" or initial tier risk assessments contribute to overestimations of exposure. Risk managers can gain important perspective by requesting quantitative disclosure of uncertainty and evaluation of conservative bias. This process can assure that conservatively-based risk estimates are appropriately qualified, or that "refined assessments" reflect realistic conditions. The Food Quality Protection Act (FQPA) requires quantitative estimation of aggregate pesticide exposure from multiple sources, routes and pathways, and where appropriate cumulative assessments must be developed for chemicals having the same mode of action. There is a continuing need to evaluate the degree of conservatism inherent in "screening-level" risk assessment methods, and thereby critique default assumptions and model uncertainty, to avoid the Precautionary Principle, and to pass the "common sense test."

Introduction

The risk assessment paradigm in use today was described in a monograph written by experts from the National Academy of Sciences (1). This paradigm recommended a clear separation of the functions of risk characterization encompassing the areas of hazard identification, dose response and exposure assessment from risk management, in which policy is delineated from the science. To appreciate the decision-making process, risk managers must understand the perspective of regulatory risk assessors in constructing the typical risk assessment (Figure 1).

A risk assessor's directive can be summarized by the Hippocratic Oath "...never do harm to anyone". Recognizing sources of uncertainty in a risk assessment, the risk assessor feels that it is imperative not to underestimate risk, i.e., if there is any error, the error must be on the side of safety (2). Even more conservative is the increasingly popular *Precautionary Principle* with the philosophy that it is better not to allow the proposed use of a chemical than risk uncertain but possibly very negative consequences (3).

In contrast to risk assessors, risk managers must deal with the political implications of the risk assessment and act to respond to the conclusions from the risk assessment. To do this, risk managers must know how much the risk may be under- or over-estimated, and balance that risk with possible benefits if the law allows it (e.g., public health protection, new tools to combat evolving resistance, etc.) or foresee real risks created by mitigating theoretical ones. They must also determine if the regulatory decision based on the risk assessment opens their agency to legal liability.

Risk Assessment

Risk Management

Figure 1. National Academy of Science Risk Assessment/Management Paradigm.

Discussion

To do their job, risk managers must be able to distinguish the possibility of risk from actual risk. Real risks are risks that will likely occur (and are known with medical certainty to have occurred previously in humans, i.e., an incidence in humans with an associated causal and/or dose-response relationship). Theoretical risks are those derived from laboratory animals given high dosages that are unlikely to occur in humans without extraordinary circumstance (high exposure). A good example is the theoretical excess cancer risk from using chlorine to purify water versus the very real risks of water-borne diseases such as cholera, poliomyelitis and a variety of disease-causing coliform bacteria. Pesticides are designed to adversely affect pests, but they may also have adverse effects on humans. To provide perspective on real risk, the risk manager should know that annually in the US, unintentional deaths from all chemicals is >14,000 persons (*4*) (See Table I). This figure does not include the >3,200 persons that die annually from drowning (*5*), nor the >5,000 persons that die each year from food-borne disease related to microbial contamination in the US (*6*). All of these mortality statistics can be considered chemical-related. And

114

Table I. Unintentional Poisoning Deaths by Chemical Class in 2001[a]

Type of Poison	All Ages
Nonopioid analgesics, antipyretics, & antirheumatics (X40)	208
Antiepileptic, sedative-hypnotic, antiparkinson & psychotrophic drugs (X41)	763
Narcotics & psychodysleptics (X42)	6,509
Other drugs acting on the autonomic nervous system (X43)	19
Other & unspecified drugs, medicaments, & biosubstances (X44)	5,525
Alcohol (45)	303
Organic solvents & halogenated hydrocarbons & their vapors (X46)	63
Other gases and vapors (X47)	593
Pesticides (X48)	7
Other & unspecified chemical and noxious substances (X49)	88
Total Poisoning Deaths	**14,078**

[a]From Reference 4.

they can be compared to the <10 deaths per year from all pesticides as a measure of real (as opposed to theoretical) risk from pesticides.

Certainly, one might argue that there is under-reporting in any of the line items in Table I. For example, Table I lists deaths from non-opioid analgesics, antipyretics and anti-rheumatics as 208, yet there are an estimated 3,200 persons that die annually from bleeding ulcers induced from non-aspirin non-steroidal anti-inflammatory drugs (7). Also, this tabulation does not contain purposefully induced death (suicide, homicide, execution, etc.). However, death (especially from acute poisoning) is an incontrovertible endpoint, and relative risk amongst causes is readily compared. While one can argue about the absolute numbers, there is generally no argument about cause of death making it a useful comparative statistic.

The historical development of the risk assessment process and its origins may be useful to the risk manager. Pesticide risk managers may not be aware that the concepts of pesticide risk assessment and risk management were adapted from the pharmaceutical regulatory process. Pharmaceutical risk managers understand that there is real risk and real benefit from using pharmaceuticals. It is not coincidence that the most prescribed drug in the US (Table II) is in the class of drugs that produce the most unintentional deaths (See Table I). What may be surprising to a pesticide risk manager is that none of the most used drugs (those with the most numbers of prescriptions written annually) in the US would meet the strict standards required for registration as pesticides. For example, of the top ten most used drugs, some have inadequate toxicity studies to be considered pesticide active ingredients, or their margins of exposure are substantially less than 100 (several ≤10), and some are *known* to produce significant adverse effects (e.g., birth defects) at therapeutic dosages in humans (Table II, derived from references *8* and *9*). There are a variety of drugs in use today that are known to produce severe birth defects in humans (thalidomide, isotretinoin, and the Measles Mumps and Rubella vaccine), and the labels on these products carry very specific warnings for women of child bearing age. Virtually every antineoplastic drug used by physicians <u>causes</u> cancer. The administration of these chemicals doesn't cause a theoretical risk, because they produce real tumors in cancer survivors (but at the same time increase life expectancy by years compared to no treatment, providing a clear example of risk/benefit). Thus, the adverse effects associated with the top ten most prescribed drugs listed in Table II are not an aberration, but reasonably represent prescription pharmaceutical drugs in general.

Table II. Top 10 Most Prescribed Drugs[a] and Their Risks[b]

Drug (Generic)	Risk of Adverse Effect
Hydrocodone	No adequate carcinogenesis studies in animals
Atorvastatin	Teratogenic in rats, rabbits; MOE = 20-30
Levothyroxine	No adequate carcinogenesis studies in animals
Atenolol	Atrial degeneration in male rats; MOE = 75
Amoxicillin	Reproductive toxicity MOE = 10
Lisinopril	Can cause fetal death in pregnant women
Hydrochlorthiazide	Reproductive toxicity MOE = 4 in rats
Furosemide	Maternal death and abortion in rabbits, MOE = 2
Albuterol	Teratogenic in mice, MOE = <1
Alprazolam	Assumed to be capable of causing congenital abnormalities

[a] From Reference *8*.

[b] Summarized from Reference *9*.

A risk manager also needs to know the type and qualities of data in a pesticide risk assessment. Of the two primary data types that go into risk assessment (toxicity and exposure), there are relatively few opportunities to refine toxicity compared to exposure data. Although there may be opportunities to refine toxicity data such as conducting a new study using more animals per dose, more dose levels, different animal strains, evaluating the mechanism of action, or examining historical controls, the exposure component of risk is much more amenable to producing new data that usually lower the risk estimate. Exposure estimates can be looked at as a tiered construct (Table III) in which the first tier is based on several defaults that tend to be quite conservative (*10*). These defaults are normally used in "screening level" or Tier I assessments, e.g., draft Registration Eligibility Decision (RED) documents that are part of the federal regulatory process. Some Tier I exposure assessments may be 10-1,000 fold higher than the actual exposure as explained later in this chapter, and there is a series of refinements (each successive refinement is more expensive to perform than the previous) that a pesticide registrant may do preemptively or be allowed to produce with concurrence of the regulatory risk manager. Examples of the series of Tiers for various pathways are shown in Table IV.

Table III: The Tiered Approach to Risk Assessment for Refining Exposure Estimates

Tier I	Series of conservative defaults
Tier II	Refine defaults with data, e.g., dermal absorption
Tier III	Stochastic analyses using full spectrum of data
Tier IV	Biomonitoring study

To illustrate the components of exposure and opportunities for refinement, we can consider the organophosphate (OP) insecticides. The EPA has produced a RED for each OP as required by the Food Quality Protection Act of 1996 and the Federal Insecticide Fungicide and Rodenticide Act as amended in 1988. These documents have prompted EPA's risk managers to request several significant reductions in OP usage that have been generally complied with by registrants of these pesticides. A brief discussion of the exposure values for dietary, drinking water, and non-dietary residential exposures abstracted from selected REDs follows. For comparison these exposure estimates were then juxtaposed with exposure estimates derived from CDC biomonitoring from the National Report on Human Exposure to Environmental Chemicals (*11*).

Each exposure pathway (food, water and residential) has its unique sources of data and input variables that are available to the risk assessor. The most refined estimates of exposure are likely for the dietary route, since both food

**Table IV: Examples of the Tiered Exposure Approach for
Various Pathways**

Pathway	Tier I: Environmental Model	Tier II: Measurement	Tier IV: Biomonitoring
Drinking Water	PRZM/EXAMS[a]	Jackson et al 2005[b]	CDC 2005[c]
Child Hand to Mouth	Smegal et al 2001[d]	ILSI 2004[e]	Krieger et al 2001[f]
Air	ISCST[g]	CDPR[h]	Osterloh et al 1989[i]
Food	Tolerances[j]	USDA PDP[k]	Curl et al 2003[l]

[a] From Reference 12.

[b] From Reference 13.

[c] From Reference 11.

[d] From Reference 14.

[e] From reference 15.

[f] From Reference 16.

[g] From Reference 17.

[h] From Reference 18.

[i] From Reference 19.

[j] From Reference 20.

[k] From Reference 21.

[l] From Reference 22.

residue data and food consumption are well-characterized by ongoing US Department of Agriculture surveys. We can compare EPA's estimated dosage from chronic dietary, chronic water and non-residential exposure derived from EPA's REDs for the most used OPs on food crops as shown in Table V. Pesticide exposure from water is notoriously exaggerated from the first and second tier models employed by EPA when compared to actual water monitoring measurements conducted by the US Geologic Survey (12). Shown in Figure 2 is the graph of predicted water concentrations from Tier I exposure assessments compared to actual measurements for 39 pesticides. There is a clear over-prediction bias. Residential exposures have even greater uncertainty, and the estimated exposures from the dermal route alone exceed the exposures estimated from all routes using biomonitoring (15) (See Figure 3).

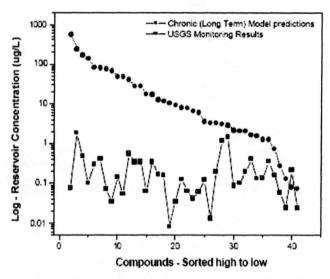

Figure 2. Pesticides in Water: Model vs. Monitoring (Reproduced from reference 12. Copyright American Chemical Society.)

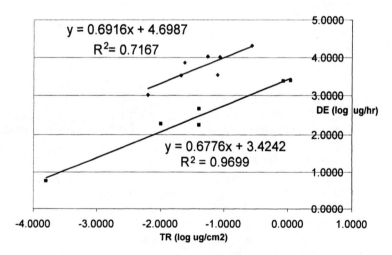

Figure 3. Regression of Log Dermal Exposure on Log Transferable Residue[a]
[a] Upper line is whole body dosimetry exposure monitoring, and lower line is from biomonitoring exposure studies (from Reference 15).

When the estimates of dosage from individual REDs are added, the total for any given pathway is greater than the estimated exposure to all OPs combined based on biomonitoring (Table V). In fact, exposure attributed to aggregated pathways from individual compounds and in some cases individual pathways for individual compounds exceed the cumulative biomonitoring total estimated to be 0.3 µg/kg bodyweight for all OPs combined. These observations should give risk managers pause to reflect on the exposure basis of calculated risk for any given pesticide. The risk manager should be asking how much conservatism (overestimation bias) is built into the risk they are about to manage based on the assessment presented to them. Interestingly, stochastic estimates of cumulative exposure tend to be very close to the estimates made from biomonitoring using alkyl phosphates (23), suggesting that it is possible to reasonably approximate true exposure with an advanced-tier analysis.

Although biomonitoring is often viewed as the "gold standard" for estimating exposure (24, 25, 26), even biomonitoring can yield an inflated estimate of total exposure (27). Biomonitoring data have been used historically as validation of passive dosimetry. However, biomonitoring extrapolated to dermal dose also tends to overestimate handler dermal exposure for two reasons (23):

- 1. The biomonitored moiety frequently represents a hydrolysis product of the parent compound and can have greater environmental persistence, allowing more contact by humans.
- 2. Biomonitoring integrates all routes of exposure including dietary, non-dietary ingestion, incidental contact, inhalation, and dermal.

We have described some of the pathways and the exposure overestimates that have been observed in them for one class of compounds (the OPs). Typically, in any given pathway, there are a number of input variables used to calculate exposure. Shown in Table VI are some of the individual input variables that contribute to overestimation bias. While some of the overestimates appear to be quite small, many of these factors are multiplicative, meaning that the overall bias is a product of multiple variables, potentially resulting in several-fold overestimates. Thus, whether occupational or residential, exposure assessors tend to overestimate exposure and the resulting risk.

Conclusions

Real risk is different from theoretical risk, although there is frequently no attempt made to distinguish between them. Theoretical risk can many times be

Table V. Dosages (µg/kg/day) Abstracted from RED Documents for Dietary, Water and Non-Dietary Residential Exposure for the Most-Used Organophosphates on Food[a]

Organophosphate	Date of EPA RED	Food Exposure	Water Exposure	Non-Dietary Exposure	Aggregate[c]
Azinphos-Methyl	05/19/99	0.195	0.125	NA[b]	0.320
Chlorpyrifos	06/08/00	0.001	0.116	0.180	0.297
Diazinon	11/14/00	0.020	0.0952	0.0036	0.119
Dimethoate	08/09/98	0.101	0.059	NA	0.160
Malathion	09/15/00	0.192	1.67	0.81	2.67
Methyl Parathion	07/28/99	0.003	0.113	NA	0.116
Methidathion	12/08/99	0.135	0.515	NA	0.650
Phosmet	09/09/99	0.022	0.0173	NA	0.039
Cumulative	NA	0.669	2.71	0.994	4.37

[a] From Reference 23.

[b] NA = Not Applicable.

[c] Sum of exposures from food, water and non-dietary.

Table VI: Sources of Overestimation Bias in Occupational and/or Residential Exposure

Variable	Tier I[a]	Realistic[b]	Overestimate
Body Weight (kg[-1])	60-70	86	1.2-1.4
Body SA (cm^2)	21,200[c]	20,700[d]	1.0
Respiration Rate (L/min)	29	9-14	2.1-3.2
Dermal Absorption (%)	100	10[e]	10
Transferable Residue (%)	5-20	0.01-12	1.7-500
Application Rate (lb/ac)	X[f]	0.5-1X	1-2
Acres Treated (ac/d)	40-1200	20-350	2-3.4
Residential Exposure Time (hr)	2-8	0.1-2	4-20
Hand to Mouth Frequency (hr[-1])	20	10	2
Hand to Mouth Surface Area (cm^2)	20	5	4
Crack and Crevice Exposure Relative to Broadcast (µg/kg)	10	2	5

[a] Exposure defaults sometimes used in a first-tier assessment.

[b] Exposure factors more physiologically or agronomically consistent with "normal".

[c] Respiration rate from Reference 28.

[d] For an 86 kg male.

[e] Average human dermal absorption of 13 different pesticides, from Reference 26.

[f] Label-specified application rate.

mitigated by simply refining the input variables to arrive at a more realistic estimate. However, the cost of taking regulatory action on the basis of excessive theoretical risk can be greater than the risk being mitigated (*29*). For example, the theoretical oncogenic risk of a vector control agent for life threatening diseases (mosquitoes carrying West Nile virus, St. Louis encephalitis or malaria) might cause the most efficacious control agent to be eliminated in favor of one that has "acceptable" oncogenic risk, but is less efficacious. A risk manager needs to consider that malaria kills >100,000 persons per year worldwide and was endemic in the US until the 1920s. Bubonic plague and its vectors are still endemic in the US today. The insect vectors for many diseases with proven body counts are only one of many examples of real risks that a risk manager must consider. Inappropriately mitigating theoretical risk can generate real risk. Risk managers have the tough job of distinguishing real from theoretical risk. William Ruckelshaus, first administrator of the US EPA, once said "Risk assessment is like a captured spy. If you torture it long enough, it will tell you what you want to know". The risk manager must distinguish what was produced under duress of a time deadline from things that really affect the quality of life. Uninformed or misinformed members of the public, public health professionals, regulators, news organizations and occasionally educators may either intentionally or unintentionally perpetuate the fear that accompanies sensationalized concern about theoretical risk. The risk manager must look at the factors underlying the "anatomy" of a quantitative risk analysis (especially default, inflated exposure estimates) and differentiate what is real versus what might be.

References

1. NAS (National Academy of Sciences) National Research Council. (1983). Risk assessment in the federal government: managing the process. National Academy Press, Washington DC.

2. US EPA (2004). An examination of EPA risk assessment principles and practices. EPA/100/B-04/001, Risk Assessment Task Force, Washinton, D.C.

3. Ricci, P.F., Cox, L.A., and MacDonald, T.R., (2004). Precautionary Priciples: A jurisdiction-free framework for decision under risk. Human and Exper. Toxicol. 23: 579-600.

4. NSC (2004). National Safety Council Injury Facts, Itasca, IL.

5. NCIPC (2005). National Center for Injury Prevention and Control http://www.cdc.gov/ncipc/factsheets/drown.htm

6. NIH (2005). National Institutes of Health, National Institute of Allergy and Infectious Diseases. Foodborne Diseases. http://www.niaid.nih.gov/factsheets/foodbornedis.htm.

7. Tarone RE, Blot WJ, McLaughlin JK. (2004). Nonselective nonaspirin nonsteroidal anti-inflammatory drugs and gastrointestinal bleeding: relative and absolute risk estimates from recent epidemiologic studies. Am J Ther. 11: 17-25.

8. Mosby (2003). Mosby's Drug Consult. Top 200 Drugs. Top 200 most prescribed drugs 2003, Elsevier, Inc. http://www.mosbysdrugconsult.com/DrugConsult/Top_200.

9. PDR (2003). Physicians Desk Reference. Thomson PDR, Montvale, NJ.

10. Krieger RI, Ross JH, Thongsinthusak T (1992). Assessing human exposures to pesticides. Rev Environ Contam Toxicol 128: 1-15.

11. CDC (2005). National Report on Human Exposure to Environmental Chemicals. http://www.cdc.gov/nceh/dls/report/.

12. Pesticide Root Zone Model (PRZM) / Exposure Analysis Modeling System (EXAMS). Available at http://www.epa.gov/oppefed1/models/water/index.htm.

13. Jackson, S., Hendley, P., Poletika, N., Jones, R., and Russell, M. (2005). Comparison of regulatory method estimated drinking water exposure concentrations with monitoring results from surface drinking water supplies J Agric Food Chem, in press.

14. Smegal, D., Dawson, J., and Evans, J. (2001). Recommended revisions to the standard operating procedures (SOPs) for residential exposure assessments. Science advisory council for exposure, policy number: 12. U.S. Environmental Protection Agency, Office of Pesticide Programs, Washington, D.C.

15. ILSI (2004). HESI Residential Exposure Factors Database Users Guide. ILSI, Health and Environmental Sciences Institute, Washington, D.C. (http://www.hesiglobal.org/Committees/TechnicalCommittees/RAM/default.htm)

16. Krieger RI, Bernard CE, Dinoff TM, Ross JH and Williams RL (2001). Biomonitoring of Persons Exposed to Insecticides Used in Residences. Ann. Occup. Hygiene, 45 (Suppl 1): 143-153.

17. Industrial Source Complex Short Term (ISCST) model, http://www.cee.odu.edu/air/isc3/odu_isc3.html.

18. California Department of Pesticide Regulation (CDPR), http://www.cdpr.ca.gov/docs/empm/pubs/tac/dichlo13.htm.

19. Osterloh, J.D., Wang, R., Schneider, F., and Maddy, K. (1989). Biological monitoring of dichloropropene: Air concentrations, urinary metabolite, and renal enzyme excretion. Arch. Environ. Health 44: 207-213.

20. Tolerances are from EPA-set limits on the amount of pesticides that may remain in or on foods, http://www.epa.gov/pesticides/regulating/tolerances.htm.

21. US Department of Agriculture Pesticide Data Program (USDA PDP), http://www.ams.usda.gov/science/pdp/.

22. Curl C, Fenske R, Elgethun K. 2003. Organophosphorous pesticide exposure of urban and suburban preschool children with organic and conventional diets. Environ Health Perspect 111:377-382.

23. Duggan, A., Charnley, G., Chen, W., Chukwudebe, A., Hawk, R., Ross, J., Krieger, R.I., and Yarborough, C. (2003). Di-alkyl phosphate biomonitoring data: Assessing cumulative exposure to organophosphate pesticides. Reg. Pharmacol. Toxicol. 37: 382-395.

24. Sexton, K., Needham, L.L., and Pirkle, J.L. (2004). Human biomonitoring of environmental chemicals. Am Scientist 92: 38-45.

25. Woollen, B. (1993). Biological monitoring for pesticide absorption. Ann. Occup. Hyg. 37: 525-540.

26. Ross, J.H., Driver, J.H., Cochran, R.C., Thongsinthusak, T. and Krieger, R.I. (2001). Could pesticide toxicology studies be more relevant to occupational risk assessment? Ann. Occup. Hygiene 45 (Suppl 1): 5-17.

27. Krieger, R., Dinoff, T., Williams, R., Zhang, X., Ross, J., Aston, L., and Myers, G. (2003). Preformed biomarkers in produce inflate human organophosphate exposure assessments. Environ. Health Persp. 111: A688-689.

28. Keigwin, T.L. (1998). PHED Surrogate Exposure Guide: Estimates of Worker Exposure From the Pesticide Handler Exposure Database Version 1.1. U.S. EPA, Office of Pesticide Programs, Washington, D.C.

29. Ricci, P.F., Cox, L.A., and MacDonald, T.R., (2005). First do no harm: Can regulatory science-policy in risk assessment be deleterious to health? Belle Newsletter, 13: 26-37.

Chapter 9

Indoor Human Pyrethrins Exposure: Contact, Absorption, Metabolism, and Urine Biomonitoring

Sami Selim[1] and Robert I. Krieger[2]

[1]Golden Pacific Laboratory, 4720 West Jennifer Avenue, Fresno, CA 93711
[2]Personal Chemical Exposure Program, Department of Entomology, University of California, Riverside, CA 92521

The exposure to people who performed structured activities (Jazzercise[TM]) on carpet treated with pyrethrins (PY), piperonyl butoxide (PBO) and N-octyl bicycloheptene dicarboximide (MGK 264) was determined. Carpets were treated with a total release fogger containing all three pesticides. Thirty subjects wore whole body dosimeters (15 subjects) or bathing suits (15 subjects) and performed the Jazzercise activity. Samples, including urine were collected and analyzed for all three compounds including the biomarker of PY, chrysanthemum dicarboxilic acid (CDCA).The deposition of the three actives on the carpets averaged 3.66, 7.26 and 10.95 $\mu g/cm^2$ for PY, PBO and MGK 264 respectively. The ratio of the mean amount of PY:PBO:MGK 264 transferred from the carpet to the whole body dosimeters was nearly indentical to the ratio of the concentration of the three compounds in the formulation indicating that the transferability is similar amongst these structurally quite different molecules. Participants in whole body dosimeters excreted 1.62 μg CDCA during a 5 day period, and those who wore swim suits excreted 13.7 μg CDCA. The calculated percent absorbed residue, from the analysis of the biomarker is comparable to the percent absorption (0.22%) calculated from a radiolabeled absorption study.

Introduction

Pyrethrum extract is a natural product derived from chrysanthemum flowers and has been used as an insecticide for centuries (Davies, 1985). The activity of this extract is predominantly due to its content of six esters, commonly referred to as pyrethrins: pyrethrin I and II, jasmolin I and II and cinerin I and II. The structures of the six esters are shown in Figure 1.

Pyrethrin I

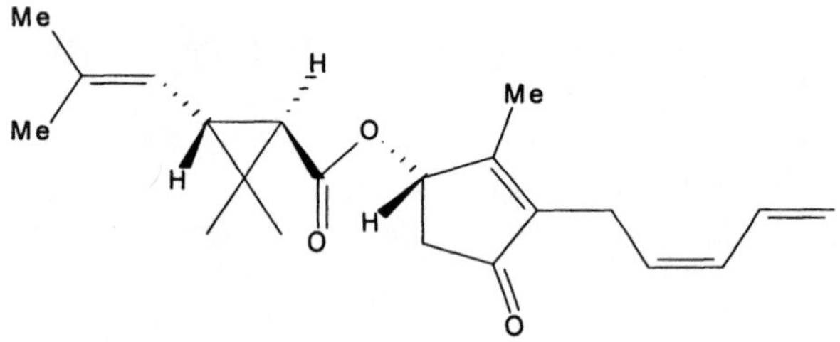

Cinerin I

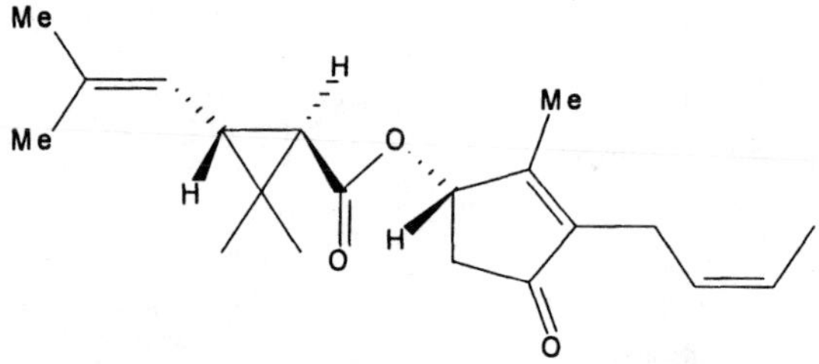

Jasmolin I

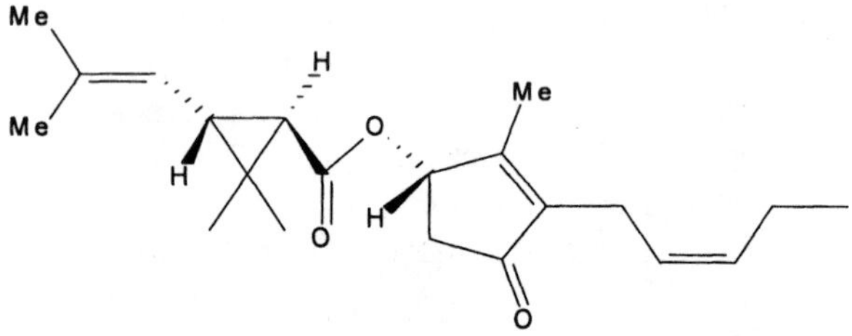

Pyrethrin II

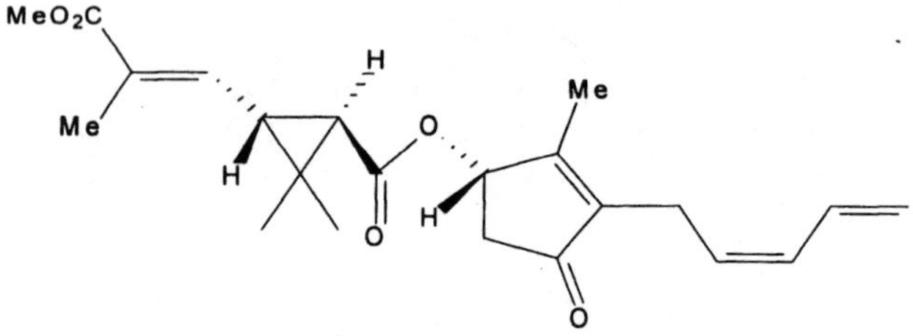

Figure 1. Structure of Natural Pyrethrins

Cinerin II

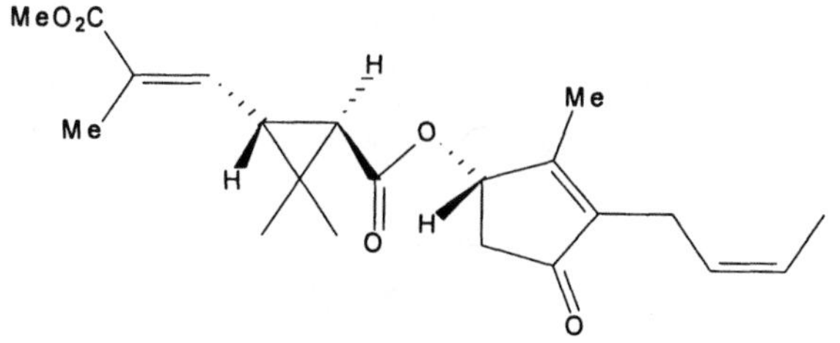

Jasmolin II

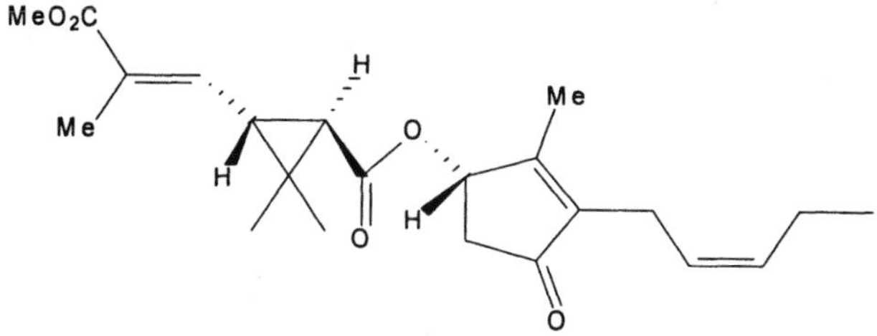

Figure 1. Continued.

Although the ratio of the 6 esters in the flowers varies, pyrethrin I which is the primary insecticidally active component of pyrethrins is the predominant ester and accounts for approximately 33% of the total.

Pyrethrins are widely used in consumer products ranging from pet care products, flying insect killers, crack and crevice treatments, as well as total release aerosol foggers to control ants, fleas, roaches, and other domestic insects. Their use can result in the deposition on various surfaces in the home. Skin contact with these surfaces can result in the transfer of low levels of residues to the skin. The nature of treated surfaces, clothing penetration, direct skin contact, and the extent of dermal absorption are among the important determinants of human exposure.

The best assessment of the safety of such exposures comes from the consideration of the absorbed dose of pyrethrins. This can commonly be estimated two ways:

1. Measuring the percentage of the dermally applied dose that crosses the skin and enters into systemic circulation. This is multiplied by the amount of residue transferred to the skin to yield absorbed dose.

2. Using a biomarker of exposure whose presence in a biological fluid is absolute evidence that the compound of interest was absorbed. This requires the specific analytical identification and quantification of the biomarker in the urine and an understanding of the conversion rate of the

parent compound to the biomarker, i.e. 100 mmoles of parent yields 52 mmole of biomarker.

There are many commercially available formulations containing pyrethrins. Piperonyl butoxide and MGK 264 synergists are used with pyrethrins in many commercial formulations (Jones, 1998). For example, piperonyl butoxide is mainly used as an adjuvant in formulation containing pyrethrins, because it delays the breakdown of pyrethrins, improves knock-down, and prolongs its effectiveness.

Biomarker Determination

An initial study was conducted to determine the biomarker for PI, the predominant ester. Once the biomarker was determined, human volunteers performed a set of structured activities (Jazzercise; Ross et al., 1990 and 1991) on carpeting which had been treated with a total release fogger containing PY, PBO and MGK 264. The activities represent extensive dermal contact, more than expected in one day.

Since the most direct procedure to determine the structure of a biomarker in urine is to use a radiolabeled compound, radiolabeled PI, the major component of pyrethrins was used in earlier studies (Selim, 2005).

A group of 3 volunteers was dosed orally with approximately 0.3 mg and 50 µCi of PI. The urine and feces samples were collected and analyzed for total radioactivity. The urine samples were also analyzed for the biomarker chrysanthemum dicarboxylic acid (CDCA) with the structure shown below:

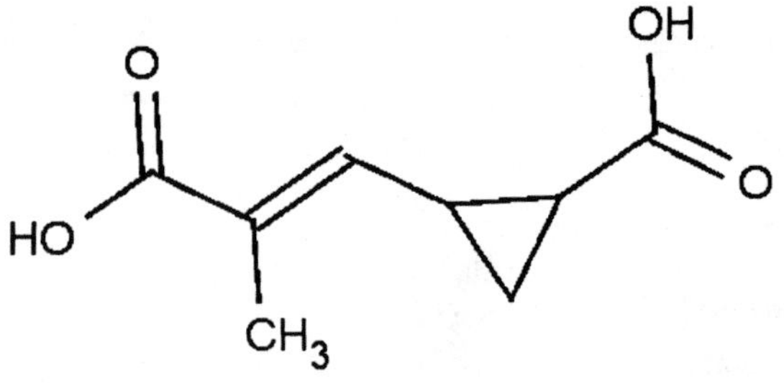

Figure 2. Structure of CDCA

The chemical structure of PI including the position of the [^{14}C] label is shown below:

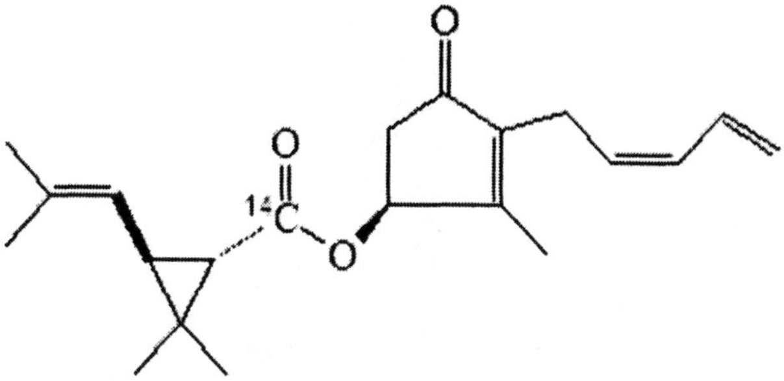

Figure 3. Structure of PI Showing Position of ^{14}C Label

The dosing solutions were prepared so that each volunteer received approximately the same total amount of PI and the same amount of radioactivity. Radiolabled compound only was used in the preparation of the dosing solution, which was prepared by adding the radiolabled compound to a 50/50 ethanol/water solution. A complete collection of urine and feces was made at 4, 8 and 12 hrs and every 12 hrs thereafter until the subjects were released from the study. Radioactivity in the urine samples was determined by transferring triplicate portions into liquid scintillation vials and counting. Fecal samples were homogenized and triplicate portions combusted in a sampler oxidizer. The radioactivity in all samples was determined by using a liquid scintillation spectrometer.

Urine samples containing sufficient radioactivity were acid hydrolyzed using hydrochloric acid. The samples were neutralized to pH of 4.0 ± 0.5 with sodium hydroxide and injected on an HPLC equipped with a Waters Symmetry C18 reverse phase column and a radioactive detector. A water (with 0.5% formic acid) acetonitrile gradient system at a flow rate of 1.0 mL/min was used to separate the radioactive peaks.

Urine samples were also analyzed by GC/MS operating in the EI mode. Urine samples were hydrolyzed, extracted and derivatized with 1,1,1,3,3,3-hexfluoro-2-propanol, prior to being injected into the GC/MS equipped with a RTX 65, 30m x 0.25m x 0.25 μm column.

Jazzercise Study

Thirty human volunteers participated in an indoor structured activity program (Ross et al., 1990). Half of the subjects wore bathing suits and the other half wore whole body dosimeters consisting of cotton socks, gloves and a union suit (long johns) (Krieger et al., 2000).

The venue was a large vacant suite in a shopping center in Riverside, California. The test room was carpeted for the study prior to the exposure period. The air exchange rate was not determined for this facility, but it was assumed to be typical of commercial retail buildings.

Five indoor total release foggers containing a water-based insecticide formulation were placed on 5 separate pieces of nylon carpet. The aerosol formulation contained pyrethrins (PY, 0.5%), piperonyl butoxide (PBO, 1%) and MGK 264 (1.67%). When activated from the center point of each carpet, the indoor fogger canister released active ingredients, along with the inerts, in a semi-directional (vertical) spray. To facilitate an even distribution of the formulation to the carpet, the total foggers were placed on a Lazy Susan, which rotated slowly during application. Alpha cellulose deposition coupons were placed at 2, 4, and 6 ft from the total release foggers to characterize the dispersion of active ingredients. Cotton percale rolled with a weighted roller (Ross et al, 1991) was used to estimate transferable residues (μg /cm^2). The product was applied following typical label instructions. The ventilation system for the test room was off for approximately 2 hours after fogger activation. After approximately 2 hours, the ventilation of the room was turned on and the alpha cellulose deposition coupons were collected. Air sampling tubes were placed above the carpet to determine the air concentration of the three active ingredients.

Following the collection of the alpha cellulose deposition coupons, the study participants were randomly assigned to spaces on the treated carpets. One group wore bathing suits and the other whole body dosimeters. Each person had a space of approximately 6 feet by 6 feet on which to exercise. A certified Jazzercise instructor led the group through a series of stretching and low impact exercises on the carpet for approximately 20 minutes (Ross et al., 1990). The 30 subjects selected for participation in the study contacted virtually every body region with the carpet during the experiment.

At the end of the Jazzercise routine, the gloves, socks, swimwear and union suits were collected. Each participant was given a supply of 4 L plastic bottles and a cooler with blue ice with instructions to collect 24 hour urine specimens each day for 5 days.

Transferability of residues to cotton percale was determined using a Modified California Roller (Ross et al., 1991, Fuller et al., 2002) as a function of a pre-specified applied force (approximately 12 kg) and contact duration (10 back-and-forth rolls).

Air-sampling was done using calibrated low-volume air-sampling pumps placed in the middle of the treated carpet. The air-flow rates were adjusted to approximately 1 L per minute. The air-sampling media consisted of sorbent passive samplers for PY, PBO and MGK 264 purchased from SKC (Fullerton, CA). There were a total of 2 air-sampling stations set up on each of the carpets designated for air-sampling. The 2 air-sampling stations on each carpet were set up at a height of 1 foot above the floor. The exact location of the sampling pumps was as close to the center of the carpet as possible (Figure 1).

The cotton gloves, cotton socks, long johns, alpha cellulose, cotton percale, and air sampling tubes were frozen until shipped to Xenos Laboratories for analysis for PY, PBO and MGK 264.

Urine samples were also frozen until shipped for analysis of the PI biomarker, CDCA.

The validated limits of quantitation (LOQ) for air sampling tubes, alpha cellulose, cotton union suits, cotton socks, cotton gloves and percale are summarized in Table I.

Table I. Limits of Quantitation

Matrix	Sample Size	Formulation mg/Sample	PYI	PY	PBO	MGK -264
			LOQ µg/Sample			
Air Sampling Tubes	1 tube	0.0200	0.0898	0.158	0.302	0.510
Alpha Cellulose	57.8 cm^2	5.00	22.5	39.5	75.5	128
Cotton Union Suits	~400g	8.00	35.9	63.1	121	204
Cotton Union Suits	~ 20g	0.400	1.80	3.16	6.04	10.2
Cotton Socks	1 sock	0.200	0.898	1.58	3.02	5.10
Cotton Gloves	1 glove	0.200	0.898	1.58	3.02	5.10
Percale	57.8 cm^2	0.100	0.499	0.789	1.51	2.55
Percale	~900 cm^2	0.744	3.34	5.87	11.2	19.0

Results

All participants completed the study and showed no signs or symptoms of adverse dermal or systemic effects from the experimental oral dose (Selim, 2005) or dermal contact with the treated carpets.

Biomarker

The percent of dosed radioactivity eliminated in urine and feces following the oral administration of radiolabled PI is shown in Table II.

Table II Mean Urinary and Fecal Excretion of ^{14}C PI Derived Radioactivity Expressed as Percent of Dose

Time Interval (hrs)	Dose in Urine (%)	Dose in Feces (%)
0-4	27.65	NA
4-8	12.00	NA
8-12	5.27	NA
12-24	6.60	4.11
24-36	2.19	NA
36-48	0.85	15.88
48-60	0.44	NA
60-72	0.23	11.38
72-84	0.16	NA
84-96	0.11	0.87
96-120	0.10	2.23
120-144	0.05	0.25
144-168	0.04	0.03
Total	55.68	34.74

Although the predominant amount of radioactivity (about 56%) is eliminated in the urine, a significant percent (about 35%) was excreted in the feces. Urinary excretion of radioactivity was rapid following oral administration. The highest percent of radioactivity was excreted in the first 4 hours after dosing (almost 40%) and declined very rapidly with a half-life of 5-6 hours.

Metabolite profiling of urine samples determined by HPLC with a radioactive detector following oral administration of ^{14}C-PI to human subjects showed the presence of two radiolabled components. The major component, which represented approximately 90% of the total radioactive chromatogram area (Table III) had a retention time similar to that of a CDCA standard (approximately 20 minutes). The minor component had a retention time of approximately 29 minutes.

Table III. Percent of Radioactivity Present as CDCA in Hydrolyzed Urine

Time Interval	% Radioactivity							
	CDCA			*Mean*	*Metabolite 2*			
	Subject Number				*Subject Number*			
	05	*06*	*07*		*05*	*06*	*07*	*Mean*
0-4	81.65	84.03	89.07	85.14	18.35	15.97	10.27	14.86
4-8	89.23	89.33	100	92.85	10.77	10.67	ND	7.15
8-12	79.88	88.10	100	89.33	20.12	11.90	ND	10.67
12-24	100	100	78.71	92.90	ND	ND	21.29	7.10
24-36	ND	ND	ND	ND	ND	ND	ND	ND
Mean				90.06				9.94

ND=Non Detected

Analysis of acid hydrolyzed urine samples by GC/MS confirmed the presence of CDCA (Leng, 2006). The concentration of CDCA in hydrolyzed urines, expressed as ng/g of urine as determined by HPLC equipped with a radioactive detector and GC/MS is shown in Table IV. The ratio of ^{14}C- CDCA determined by the two methods was calculated and found to range between 36.8 and 55.7%.

Table IV. Concentration (ng/kg) of ^{14}C CDCA Analyzed by HPLC/Radioactive Detector and GC/MS

Time	^{14}C HPLC	GC/ MS	Ratio (%)	^{14}C	GC/ MS	Ratio (%)	^{14}C	GC/ MS	Ratio (%)
0-4	227	92.8	40.9	185	103	55.7	81	40.3	49.8
4-8	84.8	35.1	41.4	96.2	49.8	51.8	68.7	34.7	50.5
8-12	27.7	14.1	50.9	24.9	13.0	52.2	9.88	4.0	40.5
12-24	15	5.5	36.8	7.39	3.0	40.6	9.0	4.8	53.3

The estimated concentrations of CDCA in urine based on the relative composition of PI-derived radioactivity were highest at the first interval following dosing (0-4 hours post-dose). Concentrations declined over time and were measurable in urine through the 12-14 hours study period.

Jazzercise

The discharge of the total release foggers simulated residential application of pyrethrins. None of the air sampling tubes contained PY, PBO or MGK 264 above the LOQ.

The mean deposition of PY, PBO and MGK 264 on the five carpets is shown in Table V. The mean deposition of the three chemicals was consistent from carpet to carpet.

Table V. PY/PBO/MGK 264 on Deposition Coupons

Carpet	Mean Deposition ($\mu g/cm^2$)		
	PY	PBO	MGK 264
1	3.64	7.53	12.4
2	3.62	7.11	11.1
3	2.93	5.96	9.24
4	4.06	8.51	11.3
5	4.07	7.18	10.7
Mean ±	3.66	7.26	10.95
SD	0.46	0.92	1.14

The deposition of all three compounds varied relative to the distance from the aerosol canister. Data for carpet 1 are shown in Figure 4.

Deposition of each of the three compounds was highest closer to the aerosol and decreased as the distance from the aerosol canister increased. The mean surface deposition levels ($\mu g/cm^2$) for all 3 compounds is shown in Table VI. The PY surface level represents the source of contact-transfer dermal exposure under normal conditions of use. When a transfer coefficient is established (cm^2/h or cm^2/day), the potential dermal exposure can be estimated for product development or regulatory purposes.

The total amount of PY, PBO and MGK 264 transferred from the carpet to percale using the roller at a distance of 2 ft or 6 ft from the aerosol canister is shown below (Table VII).

Deposition levels on alpha cellulose were proportional to the composition of the test material. The percent of compound transferred from the carpet surface to percale was divided by the amount of compound on alpha cellulose pads. It clearly showed that transfer was not affected by the structure of the

Table VI. Surface Deposition of Fogger Constituents

Compound	Mean Deposition ($\mu g/cm^2$)	Ratio	Ratio in Formulation
PY	3.66	1	1
PBO	7.26	1.98	1.9
MGK 264	10.95	2.99	3.23

compound or by the amount of compound on the carpet. The residues (μg) of all three compounds in cotton suits, cotton socks and cotton gloves is shown in Table VIII. These residues represent potential dermal exposure (PDE; μg/person). The respective PDE levels were similar to deposition levels of the 3 analytes. The ratios of PY:PBO:MGK 264 was about 1:2:3 in both cases (Tables VI and VIII) and very similar to the fogger formulation (Table VI).

Table VII. Surface Deposition as a Function of Distance from the Fogger

Compound	2 ft		6 ft	
	Deposition on Alpha Cellulose ($\mu g/cm^2$)	Percent Transfer	Deposition on Alpha Cellulose ($\mu g/cm^2$)	Percent Transfer
PY	6.68	8.69	1.51	7.14
PBO	13.06	7.66	2.91	7.56
MGK 264	19.06	8.32	4.56	6.48

Table VIII. Residues in Whole Body Dosimeter Matrices

Matrix	Total/ Unit (μg/person)		
	PY	PBO	MGK 264
Cotton Suits	3709	7635	12691
Cotton Socks	1602	2853	4302
Cotton Gloves	1009	1967	2904
Total	6320	12455	19897
Ratio Actual (Table 5)	1.0	2.0	3.3
Measured Ratio	1	1.97	3.1

			Deposition μg/cm²	Distance from Canister
CARPET		PY	2.14	6 ft
		PBO	4.20	6 ft
		MGK	7.75	6 ft
		PY	3.47	4 ft
		PBO	7.09	4 ft
		MGK	12.5	4 ft
		PY	5.11	2 ft
		PBO	11.2	2 ft
		MGK	17.4	2 ft
		PY	7.51	2 ft
		PBO	15.7	2 ft
		MGK	25.3	2 ft
		PY	2.57	4 ft
		PBO	5.14	4 ft
		MGK	8.37	4 ft
		PY	1.07	6 ft
		PBO	1.89	6 ft
		MGK	3.34	6 ft

☐ Alpha Cellulose Deposition Coupon   ● Aerosol Canister

Figure 4. Deposition of PY/PBO/MGK 264 (μg/cm²) on Alpha Cellulose for First Carpet

The total mean μg CDCA eliminated in urine of participants wearing whole body suit or swim wear is shown below in Table IX.

Skin exposure of persons who wore swimwear was 9-times greater than exposures of persons who wore whole body dosimeters. Assuming that all 6 isomers of PY are metabolized to CDCA and 56% of orally dosed PI is excreted in the urine, and that about 90% of the excreted compound is CDCA after correcting for molecule weight, the mean urine excretion represents an absorbed dose of about 47 μg of PY.

Absorption and elimination of PY by persons wearing whole body dosimeter was substantially less than that of the high skin exposure persons. The swim wear group eliminated 13.7 μg CDCA over the 5-day monitoring period. The whole

Table IX. Urine Biomonitoring Following Jazzercise Exposure on Treated Carpet

Day	Mean μg of CDCA Excreted in Urine	
	Swim Wear	Whole Body Suit
1	6.2	0.50
2	4.3	0.36
3	1.9	0.39
4	0.8	0.18
5	0.5	0.19
Total	**13.7**	**1.62**

body suit group eliminated only 1.6 μg CDCA during the same period. The CDCA urine elimination half-life was 24 hours. On this basis the monitoring period was sufficient to collect nearly all of the dose. The environmental and biological samples can be used to evaluate the dependence of absorbed dose on environmental residues (Table X). Aerosol deposition samples collected on α–cellulose coupons contained a mean of 3.7 μg PY/cm². The transferable residue collected on percale rolled with a weighted cylinder (Ross et al., 1990) was 0.34 μg /cm². The available residue was sampled using cotton whole body dosimeters and assuming a body surface area of 20,000 cm². The mean available PY residues on union suits, cotton socks and cotton gloves was 3709 μg, 1602 μg and 1009 μg , respectively for a total of 6320 μg. The absorbed dose was 47 μg completing the mass balance. From these data dermal absorption (47/6320 x 100 = 0.74%) and a transfer coefficient [TC =Available residue/ (Transferable residue x day)] of TC = 6320 μg /0.34 μg /cm² x day = 18,000cm²/day. Day rather than minutes or hours is used as time factor due to intensive exposures of the structured activity. This is consistent with the highly variable nature of indoor activities (as opposed to a highly stereotypic work task such as harvesting field crops) and unpublished pyrethroid monitoring studies that have been performed at the University of California, Riverside (Krieger et al., unpublished)

Conclusion

A biomarker was identified following oral administration of ^{14}C PI. The principal route of excretion of radioactivity was via the urine, but excretion via the feces was significant. This is similar to what was seen with the rats. (Selim, 1995). HPLC analysis of acid hydrolyzed urine showed that CDCA was the predominant biomarker (Figure 2). This was confirmed by quantitatively comparing the amount of radioactive metabolite in urine by HPLC/radioactive detector and by GC/MS. The conversion rate from ^{14}C PI urinary derived radioactivity to CDCA ranged from 78-100% with an overall mean of 90%. This

Table X. Total and Transferable Surface Residues of Pyrethrins

Exposure Metric	Matrix		Measure
Fogger deposition	Total surface residue $\mu g/cm^2$ α-cellulose		$3.7\ \mu g/cm^2$
Transferable residue	Rolled cotton $\mu g/cm^2$ percale		$0.34\ \mu g/cm^2$
Available residue	Whole body dosimeter μg /person	Gloves Socks Union suit	1009 ± 377 1602 ± 723 3709 ± 2196
		Total availble	$6320\ \mu g$ /day
	Mean available residue	$6320\ \mu g/20000$ cm^2	$0.32\ \mu g/cm^2$
	Clothing penetration	$0.32/3.7 \times 100$	8.6%
Available dosimeter penetrated residue urine μg/person		μg PY=1.62 μg CDCA x 1.93	$3.1\ \mu g$ PY
Absorbed Dose PY		13.7 μg CDCA x 1.93/0.56	$47\ \mu g$
Dermal absorption	Absorbed dose/ Available residue	$47\ \mu g$ /(6320 + 47) μg	0.74%
	Potential dose (PD)	$3.1 \times 100/0.74$	$419\ \mu g$
	Dose = PD / Total available + AD	$419\ \mu g$ /7095 μg	5.9%
Absorbed Daily PY Dosage		$47\ \mu g$ /70 kg	$0.67\ \mu g$ /kg-day
Estimated total available PY residue		$(100/0.74)(47\ \mu g$ /person)	$6351\ \mu g$

critical data were used in the analysis of the human disposition of pyrethrins in commercial foggers.

This study provides important insight into the transfer of pesticide residues on carpet to humans. For the whole body dosimeter as a whole, as well as for the individual components (suit, gloves, and socks) considered separately, the ratio of the mean amount transfered for PY:PBO:MGK-264 was nearly identical to the ratio of the concentration of the three compounds in the formulation indicating that the transferability is similar amongst these structurally quite different molecules. This observation will assist in the evaluation of the exposure potential of related products. Similarly, comparable percent transferabilities were seen with PY, PBO, and MGK-264 in the transferability studies conducted with the roller and cotton cloth.

In the present study, based on air monitoring conducted during the choreographed activity, inhalation exposure should essentially be zero.

References

Davies, J. H. (1985). The pyrethroids: An historical introduction. In The Pyrethroid Insecticides, ed J. P. Leahy, Taylor & Francis, London. 1-41

Fuller, R., Klonne, D., Rosenheck, L., Eberhart, D., Worgan, J., Ross, J. (2001) Modified California Roller for Measuring Transferable Residues on Treated Turfgrass. Bull Environ Contam Toxicol 67:787-794.

Jones, D. G. ed. (1998). Piperonyl Butoxide, Academic Press, San Diego, CA. 323 pp.

Krieger, R. I., Bernard, C. E. , Dinoff, T. M., Fell, L., Osimitz, T. G., Ross, J. H. and Thongsinthusak, T. (2000). Biomonitoring and whole body dosimetry to estimate potential human dermal exposure to semivolatile chemicals. J. Exposure Anal. Environ. Epidemiol. 10: 50-57.

Leng, G., Gries, W. and Selim, S (2006) Biomarker of Pyrethrum Exposure. Toxicology Letters 162:195-201.

Ross, J., Thongsinthusak, T., Fong, H. R., Margetich, S. and Krieger, R. (1990). Measuring potential dermal transfer of surface pesticide residue generated from indoor fogger use. An interim report. Chemosphere 20: 349-360.

Ross, J., Fong, H. R., Thongsinthusak, T., Margetich, S. and Krieger, R. (1991). Measuring potential dermal transfer of surface pesticide residue generated from indoor fogger use. Interim report II. Chemosphere 22: 975-984.

Selim,S. (1995) Pharmacokinetics and metabolism of pyrethrin I in the rat. BTC Study P1092006.

Selim, S. (2005) A Single Dose, Open Label Study to Investigate the Absorption and Excretion of Orally Administered or Dermally Applied [^{14}C]-Labeled Pyrethrin I (PI) to Healthy Male Volunteers. Study SEL 0204.

Williams, R. L., Bernard, C. E. and Krieger, R. I. (2000). Human exposure to indoor residential cyflythrin residues during a structured activity program. J. Exposure Anal. Environ. Epidemiol. 13: 112-119.

Williams, R. L., Oliver, M. R., Ries, S. B., and Krieger, R. I. (2003). Transferable chlorpyrifos residue from turf grass and an empirical transfer coefficient for human exposure assessments. Bull Environ Contam Toxicol. 70: 644-651.

Chapter 10

Monitoring Human Exposure to Pesticides Using Immunoassay

Marja E. Koivunen[1,2], Shirley J. Gee[1], Mikaela Nichkova[1], Ki Chang Ahn[1], and Bruce D. Hammock[1]

[1]Department of Entomology, University of California, Davis, CA 95616
[2]Current address: Antibodies Incorporated, P.O. Box 1560, Davis, CA 95617–1560

Immunoassays offer selective, sensitive and low-cost tools for the assessment of pesticide exposure through the measurement of parent compounds or key metabolites in biological fluids such as urine, blood or saliva. Recently conducted biomonitoring studies for paraquat and atrazine illustrate the strengths and weaknesses of the immunochemical approach. Development for improved assay throughput and sensitivity includes substitution of enzyme labels with fluorescent nanoparticle probes or luminescent acridinium labels together with the use of automated immunoanalyzers, immunosensors or microchips with flow-through systems.

Introduction

Biological monitoring of exposure is currently applied in environmental and occupational toxicology as well as in epidemiological studies on the dose-response relationship between internal exposure and adverse health effects. All three types of biomarkers – exposure, effect and susceptibility – can be used for pesticide exposure assessment. However, the biomarkers most often used in pesticide studies, biomarkers of exposure, are the ones indicating recent or long-term exposure to a particular compound of interest (1) .

Chromatographic techniques used for exposure analysis (2,3) are often expensive requiring special instruments and extensive sample clean-up, extraction or derivatization. Modern immunochemical techniques offer simple, specific and sensitive tools for human exposure studies involving numerous samples in complex matrices (4).

Selection of Biomarkers and Sample Media

Biomarker

Overall, the choice of a biomarker for a particular compound requires extensive knowledge about its biotransformation and metabolism in humans. Studies on animals give only a partial answer to this question, and in some cases, a compound identified as a major metabolite in a high-dose animal study is not the same in humans exposed to concentrations relevant for occupational and nonoccupational settings (5). It has been suggested that in order to properly evaluate biomarkers of exposure to pesticides, human volunteers should be given low doses of the compound (6). Similarly, measurement of a parent compound or its main metabolites can rarely provide any information on the health risk unless it is tied into an epidemiological study with corresponding short and long-term health effect evaluations.

According to Hoet (7), biological monitoring of exposure to pesticides is aimed at the estimation of internal dose based on the fate of the compound in human body. Biological monitoring approaches can be categorized into four main types

1. Direct measurement of unchanged pesticides in biological matrices (2,4-D, pentachlorophenol, DDT, lindane, paraquat)
2. Determination of metabolites in biological matrices (atrazine mercapturate, 3-phenoxybenzoic acid (PBA), *cis/trans*-dichlorovinylcyclopropane acid (DCCA), 1-naphtol)
3. Quantification of biological effects related to internal dose (acetylcholinesterase activity)
4. Measurement of macromolecule adducts combined with target or non-target molecules (DNA and hemoglobin adducts)

Biological medium

The selection of biological medium for monitoring is determined based on the excretion pattern of the selected analyte as well as the ease and timing of sampling and the availability of data relating excretion to exposure. Depending on the analyte, either xenobiotic parent compounds or metabolites can be analyzed in urine, which is by far the matrix most often used for biological monitoring of human exposure. Two obvious advantages of urine over blood are its ease of availability and the amount of sample available for analysis. In some cases, the concentration of toxicants or metabolites is higher in urine than in blood, which decreases the sensitivity requirement for the analysis method.

Blood is an excellent medium for biological monitoring but its sampling requires an invasive procedure which makes it less desirable for routine analysis in the field. For immunochemical methods, serum and plasma are preferred matrices over whole blood and ELISAs have been successfully applied in the analysis of human plasma for phenylurea (8) and triazine herbicides (9).The blood concentration of the parent compound is usually highest immediately following exposure, and the preferred time for sampling is easy to establish. However, the blood volume obtained is usually small, which means that ultrasensitive analytical techniques might be required. It should be kept in mind that an increase in sample size is usually accompanied by an increase in background noise. With immunochemical detection methods, this usually means that in order to remove the interfering substances from the sample matrix, sample preparation steps are needed.

Besides urine and blood, many other biological media are available for sampling. Hair, nails, saliva, milk, feces and fat tissue can be used for biomonitoring, but for all these matrices, tedious sample pretreatment and extraction methods are usually required prior to analysis. A few studies have used ELISA for the analysis of pesticide residue in saliva: Denovan et al. (10) analyzed atrazine in human saliva samples using ELISA, and concluded that the salivary concentration of atrazine was a good indicator for human exposure to this pesticide. The assay had a limit of detection of 0.22 ng mL^{-1} in saliva and could clearly distinguish among workers who had sprayed atrazine. Up to date, strong correlation with plasma and saliva concentrations has only been demonstrated in animal models for atrazine (11) and diazinon (12). More studies on the relationship between human plasma and saliva concentrations of environmental contaminants are needed in order to fully utilize saliva biomonitoring for estimation of absorption, metabolism and excretion.

Immunoassays for biomonitoring

In the past 10 years, the application of immunoanalytical techniques in human exposure studies has steadily increased. However, most field studies have been targeting human exposure to industrial chemicals and their metabolites (13-16). A review by Barr and Needham (3) listed only a few papers reporting use of immunochemistry in the analysis of pesticides or their metabolites in biological matrices. Since 2002, a few successful attempts have been made in our laboratory (17,18) to use these rapid, simple and cost-effective methods for human biomonitoring of pesticide exposure. The recent rapid growth of immunoanalytical techniques in clinical chemistry can be attributed to the low cost and portability of assays – simple, yet sensitive immunoassays can be performed without extensive training, even in the field and in the point-of-care facilities. The increased availability of polyclonal, monoclonal and even recombinant antibodies for a variety of pesticides and their metabolites within the academia has contributed to the growth of immunochemical analysis in the field of human exposure monitoring. However, the availability of low-cost analysis kits and reagents still does not meet the increasing demand, which limits the application of immunoassays in the field studies. Another factor making immunoassays more applicable to human biomonitoring studies is the substitution of enzyme labels by fluorescent probes, which efficiently reduces background signals and enhances the limit of detection (LOD). Schobel et al. (19) have given a comprehensive review on immunoanalytical techniques and fluorescence detection suitable for pesticide residue monitoring in environmental and food analysis. Similarly, development of fluorescent probes and immunobiosensors is gradually advancing the technology from the conventional competitive microplate pesticide assays to a more efficient and sensitive detection of multiple analytes in complex biological matrices (20,21).

Special Requirements

Elimination of Interfering Substances

In biological samples, the interferences caused by matrix components can vary from sample to sample, and endogenous compounds with structural similarities with the target analyte can interfere with the immunochemical detection. In some cases, a simple sample dilution is enough to eliminate interfering substances for ELISA analysis (22,23). However, this might decrease the assay sensitivity and increase the limit of detection (LOD). Therefore, a basic requirement for sample dilution is high enough assay sensitivity, which allows detection at low analyte concentration. Biagini et al. (20,21) used a 1:10 dilution before the analysis of pesticide residues in urine by multiplexed fluorescence microbead covalent assay (FCMIA). For the ELISA analysis of metolachlor (24) and atrazine mercapturate (25) a dilution factor greater than 1:10 was needed to eliminate the interfering substances in the urine matrix. Urine samples analyzed for a chlorpyrifos metabolite, 3,5,6 trichloro-2-pyridinol (TCP) (26), and atrazine (27) required at least a 1:50 dilution. Elimination of the urine matrix effect required a substantial sample dilution (1:100 to > 1:1000) in the ELISAs for pyrethroid metabolites (18). Instead of dilution, Lyubimov et al. (28) took a simplified approach of using the ratio between 2,4-D-spiked and non-spiked samples to minimize the effect of interfering substances in urine.

For some assays, though, a more complete clean-up and elimination of interfering substances are required (13, 15, 29, 30). Since immunoassays tend to tolerate up to about 20 % of many organic solvents, the conventional sample preparation techniques like liquid-liquid extraction (LLE), and solid-phase extraction (SPE) can be easily coupled to immunoassays. SPE with C-18 reversed phase resin has been the method of choice in studies on urinary biomarkers (*31, 32*). In our studies on paraquat (17) and atrazine mercapturate (30) in human urine samples, SPE cleanup with a mixed mode strong cation exchange resin (Oasis-MCX) was required before ELISA. In some cases, immunoaffinity chromatography can be used for sample clean-up and concentration before immunoassay as demonstrated by Nichkova and Marco (33).

For blood analysis, only minimal sample preparation has been used prior to immunochemical analysis. For example, a competitive ELISA for the phenylurea herbicide chlortoluron (8) in human plasma required no sample pretreatment. Önnerfjord et al. (9) used a flow immunoassay with fluorescence detection for the measurement of triazine herbicides in human urine and plasma. For urine, a simple dilution was a sufficient pretreatment step but for the plasma samples, a SPE clean up with a restricted access (RA) C-18 column was needed in order to make the system more sensitive. In the study by Denovan et al. (10), saliva clean-up with solid phase extraction (SPE) using C-18 cartridges was necessary in order to minimize the saliva matrix effects detected previously (11).

Specific Examples from UC Davis

Assays for Insecticide Metabolites – Pyrethroid Metabolites

Pyrethroids are highly potent insecticides that have been widely used in agriculture, forestry, horticulture, animal and public health, and in households (34, 35). Out of all pyrethroids, the most common one, permethrin, is also used as the active ingredient in personal care products, such as shampoos and lotions for lice (36). Although pyrethroids are considered safe for humans because of their relatively low mammalian toxicity, numerous studies have shown that very high exposure to them might cause potential problems such as suppressive effects on the immune system, endocrine disruption (37), lymph node and splenic damage, and carcinogenesis (38). Kolaczinski and Curtis (39) have recently reviewed the debate on chronic illness as a result of low-level exposure to synthetic pyrethroid insecticides.

In mammals, pyrethroids are metabolized rapidly by oxidation and hydrolytic cleavage of the ester linkage, followed by various species-dependent conjugations such as to glucuronide, glycine, taurine, and sulfate (40,41) (Figure 1). Although no study has specifically determined the nature of the conjugates of pyrethroids in humans, it has been well established that glycine is the most common amino acid used in conjugation reactions with xenobiotics containing a carboxylic acid group (42). Thus, the effort of some researchers in our laboratory has been focused on the development of immunoassays for the glycine conjugates of the respective pyrethroid metabolites, for example immunoassays for s-fenvalerate acid (sFA)-glycine as a biomarker for esfenvalerate exposure (32) and for the glycine conjugate of *cis/trans*-3-(2,2-dichlorovinyl)-2,2-dimethylcyclopropane (DCCA), the major metabolite of permethrin (43).

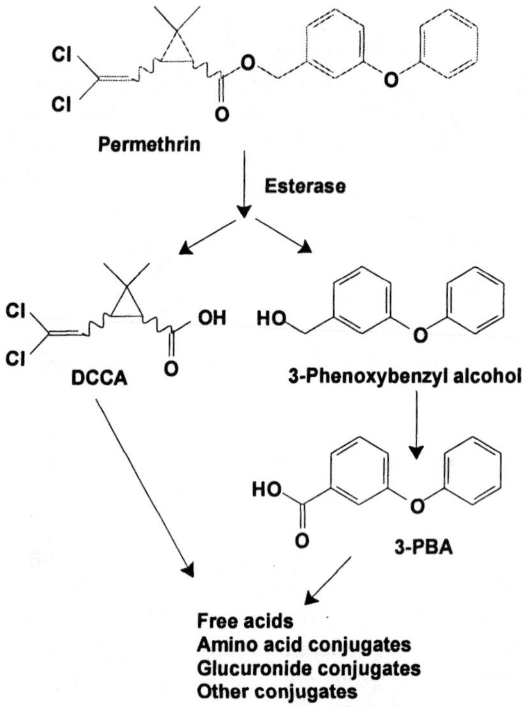

Figure 1. Metabolism of permethrin in mammals (40, 41)

The competitive indirect ELISAs for the detection of esfenvalerate metabolites (sFA-glycine and PBA-glycine) in human urine feature linear ranges for the optimized standard curves of approximately 0.03-60 ng mL^{-1} and 0.04-50 ng mL^{-1}, respectively. Both immunoassays are highly specific, and the lack of crossreactivity with the free PBA and FA makes these assays very useful for selective detection of esfenvalerate metabolites. The ELISAs were applied to the quantitative detection of trace amounts of sFA- and PBA-glycines in human urine by using either direct dilution (150-fold) or a C18-SPE method. The introduction of the C18-SPE followed by 50-fold dilution allowed a limit of quantification (LOQ) of 1 ng mL^{-1} of sFA- and PBA-glycines. Both assays were validated in a blind fashion for 15 urine samples from individuals with no known exposure to pyrethorids, and an excellent correlation between spiked and measured concentration by the ELISAs was observed (32).

Another type of immunoassay, homogeneous fluoroimmunoassay for PBA-glycine, using the polyclonal antisera has also been reported (44). This quenching fluoroimmunoassay (QFIA) is based on the competition between a fluorescein-labeled and a non-labeled glycine conjugate of PBA. The assay offers an LOD of 0.25 ng mL^{-1} in buffer. Background fluorescence from urine samples was eliminated by 1000-fold sample dilution, which limits the applicability of this method to the detection of pyrethroid metabolites in pest control operators who may have been highly exposed. The average analytical recovery obtained for 12 spiked urine samples was 85-111%. Several factors, such as the lack of washing steps, short incubation of the immunoreagents (25 min) and the fast measurement (5 s) make the assay attractive for a rapid screening method to separate samples that can be further analyzed by more sensitive instrumental or ELISA methods.

The toxicity of the insecticide permethrin is dependent on its three-dimensional configuration. The *cis*-isomer is more toxic than the *trans*-isomer. However, *trans*-permethrin predominates (60-75%) in the commercial product, and the amount of free *trans*-DCCA in human urine ranges from 65 to 87% (45,46). With the aim to detect the *cis/trans*- DCCA metabolites several sensitive ELISAs with a heterologous configuration (*cis/trans* and *trans/cis*) between antibody specificity and hapten structure of the coating antigen have been reported recently (43). The IC50 values are as low as 1.3-2.2 ng mL^{-1} for *trans*- DCCA-glycine and 0.4-2.8 ng mL^{-1} for *cis*- DCCA-glycine in buffer. Among these assays the best combination for the detection of *cis/trans*- DCCA-glycine has been chosen for further optimization and application to urine samples (47). The quantitative and sensitive detection of DCCA-glycine by the ELISA in urine samples was achieved after C18-SPE clean up and further 5-fold dilution that completely removed the urine matrix interferences. This method has a LOD of 1 ng mL^{-1}. The method was validated by the ELISA analysis of urine samples from 12 non-exposed individuals spiked with a mixture of *cis/trans*- DCCA-glycine (40:60), and very good correlation between spiked and measured was observed (R^2=0.98).

Since most pyrethroids contain the phenoxybenzyl group, monitoring the common metabolite, 3-PBA, in urine would allow the evaluation of the human exposure to all pyrethroids containing this moiety. This was the objective of the development of a 3-PBA immunoassay by Shan et al. (18). This competitive ELISA obtained had a dynamic range of 0.1-5 ng mL^{-1} with an IC50 value of 1.65 ng mL^{-1} 3-PBA in buffer, which compares well with chromatographic methods reviewed by Aprea et al. (2). The 3-PBA immunoassay is highly specific for the target analyte PBA and the related cyfluthrin metabolite (4-fluoro-3-phenoxybenzoic acid). The crossreactivity with parent pyrethroids and other metabolites is negligible. Urine matrix effects were eliminated by a simple 100-fold dilution prior to ELISA and the linear regression analysis of ELISA results of spiked urine samples from non-exposed people showed a good correlation (R^2=0.900). Furthermore, a good correlation between ELISA and GC-MS values was achieved in samples from exposed workers suggesting that the PBA immunoassay is useful for human exposure monitoring and toxicological studies.

Assays for Herbicides – Paraquat

Paraquat (1,1'dimethyl-4,4' bipyridinium) is a fast-acting, quaternary ammonium, non-selective, contact herbicide, which inhibits photosynthesis when applied to plant foliage. It is used extensively for both weed control and as a pre-harvest desiccant and defoliant. Although paraquat is highly water soluble, it is not easily leached from soil or taken up into plant root systems as it is quickly and strongly adsorbed to clay and soil organic matter. The extensive research on the fate of paraquat in agroecosystems has been reviewed by Summers (48) and recently by Roberts et al. (49).

Determination of the paraquat concentration in urine is a valuable tool for diagnosis in accidental, suicidal, and occupational intoxications, and it can be used for biological exposure assessment as well. In mid 1980's, Van Emon et al. (50) developed a competitive enzyme-linked immunosorbent assay (ELISA) for measurement of paraquat in human exposure samples. In comparison with a gas chromatographic method, the ELISA gave higher recoveries, was less labor intensive, and was more sensitive (LOD 0.1-1.0 ng mL^{-1}). This same method was recently used for paraquat analysis in an epidemiological study conducted by UCD in 2001-2003 (17).

This SALUD (Study of Agricultural Lung Disease) study tested the hypothesis that a low-level paraquat exposure can have adverse health effects including restrictive lung function. The study population included a total of 338 farm workers in Costa Rica, both pesticide handlers and non-handlers. In the same study, the paraquat ELISA was also used for measurement of paraquat trapped in air filters simulating the potential for exposure through inhalation. Prior to analysis, interfering substances in the urine samples were removed using Oasis-MCX (mixed mode cation exchange resin) SPE. When the results obtained by ELISA were validated against a current LC-MS method, the correlation between results for blind samples obtained using ELISA and LC-MS was significant (R^2 = 0.945 and 0.906 for spiked and field samples, respectively). This ELISA method had a limit of quantification of 2 ng mL^{-1} which is 5-fold lower than obtained with LC/MS/MS methods published recently (51). The paraquat ELISA was able to distinguish farm workers who were exposed from those non-exposed (Table 1). For comparison, for the air filter analysis, paraquat was first extracted by 9 M H_2SO_4 at 60 °C for 12 hours, and the results obtained by ELISA showed good correlation (R^2 = 0.918) with the UV (256 nm) measurements (17).

Table I. Amount of paraquat excreted in urine (µg 24 h^{-1}) in samples analyzed by ELISA

Group	Number of Samples	Paraquat (µg 24 h^{-1})	Range (µg 24 h^{-1})	% of Samples lower than LOQ
Control 1	30	-	-	100.0
Control 2	53	0.31	0 – 6.8	92.5
Handler	119	5.64	0 – 75.4	47.0

NOTE:　Control 1: Farm workers on control farms where no paraquat was used
　　　　Control 2: Farm workers not handling paraquat on farms where it was used
　　　　Handler: Farm workers who handle paraquat on farms where it was used
SOURCE: Data are from reference 17

Assays for Herbicide Metabolites - Atrazine Mercapturate

Atrazine is one of the most widely used herbicides in the United States. Due to its fairly good mobility in soil, it is one of the main surface water contaminants in the Midwestern United States (52,53). Atrazine has a low toxicity to humans but it has been implicated as a clastogen, an agent that causes chromosomal damage (54) and quite recently, atrazine at environmental concentrations has been found to have adverse effects on the development of anuran larvae (55). Despite its low acute toxicity to humans, atrazine is a potential endocrine disrupter and possible carcinogen, which poses a health risk to humans, especially to agricultural workers through occupational exposure.

Due to its rapid detoxification, atrazine metabolites are more likely to be found in urine and feces than the parent compound (56,57). The main urinary metabolite, atrazine mercapturate, (N-acetyl cysteine derivative of atrazine) (22) is quite stable and hence, can be used as a biomarker for atrazine exposure in humans (58). The relevance of urinary atrazine mercapturate (AM) in human metabolism has been confirmed using high-performance liquid chromatography–accelerator mass spectrometry (HPLC-AMS) to detect urinary atrazine metabolites after a dermal exposure to ^{14}C-labeled atrazine (58).

An enzyme-linked immunosorbent assay (ELISA) for AM was first developed in our laboratory by Lucas et al. (22). The assay was based on a monoclonal antibody, and was able to detect AM down to 0.5 ng mL^{-1} in crude urine diluted to 25 % with buffer. Similarly, an ELISA based on a polyclonal anti-AM antibody offered a limit of quantification of 0.3 ng mL^{-1} after a simple (1:4) sample dilution (23). Because the levels of urinary metabolites measured in epidemiological studies are usually quite low, high assay sensitivity is often required. We attempted to improve the sensitivity of the atrazine mercapturate-ELISA using different SPE pretreatment methods, and compared them with sample dilution (30). For data validation, a new HPLC/MS method using on-line SPE and column switching was also developed for the analysis of atrazine mercapturate in human urine samples. Methods were further assessed and validated using a set of field urine samples collected in the National Cancer Institute (NCI) corn farming study.

Of the two SPE resins tested, the mixed-mode resin Oasis-MCX was more compatible with immunochemical analysis than the reversed-phase Oasis HLB. On the other hand, the HLB resin performed well as an HPLC precolumn. Obviously, atrazine mercapturate containing both hydrophobic and hydrophilic moieties is a challenging analyte for solid phase extraction. This finding indicates that for analytes like AM, a resin suitable for an LC/MS online SPE does not necessarily offer the best separation for ELISA, which requires more complete elimination of structurally related, interfering substances.

Validation of all three methods, LC-MS, ELISA+SPE, and ELISA+sample dilution with spiked urine samples showed good correlation between the known and measured concentrations with R^2 values of 0.996, 0.957 and 0.961, respectively (Figure 2). Overall, both ELISA methods tended to underestimate the urinary AM concentration (slopes = 0.84-0.86), which is probably an indication of an incomplete recovery of AM during the SPE clean-up. When a set (n=70) of urine samples from a corn farming study was analyzed, there was a good agreement (R^2 = 0.917) between the ln-transformed values obtained by ELISA+SPE and LC-MS suggesting that both methods would be suitable for the analysis of urinary AM as a biomarker for human exposure of atrazine. Both methods have similar limits of detection (SPE+ELISA 0.04 ng mL^{-1} with a 1-mL sample, LC/MS 0.05 ng mL^{-1} with a 10-μL sample), which are 5-10 fold lower than the ones previously reported in the literature (22,23,59).

Future directions

As already mentioned, the development of sensitive immunoassays for human biomonitoring has been quite intense during the past decade. However, most field studies using immunochemical analysis have been targeting industrial chemicals, not pesticides. One of the limiting factors for the analysis of pesticide residues in large-scale field studies is probably the lack of commercially available, affordable immunoreagents and kits for pesticide analysis.

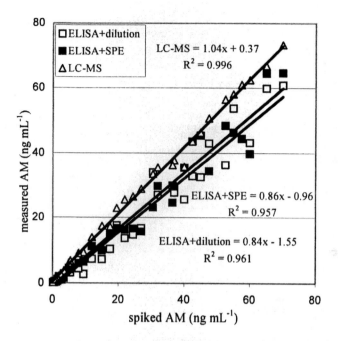

Figure 2. Correlation between known and measured concentrations of AM in urine samples obtained by ELISA or LC-MS. Data obtained from ref 30.

Another factor limiting the use of immunoassays in field studies is the low throughput of regular 96-well plate assays. In order to increase both throughput and sensitivity of immunochemical analysis, a concept of chemiluminescent immunoassay has been tested in our laboratory (60). The system is based on a chemiluminescent reporter, acridinium, which gives a detectable signal in less than 2 seconds. Acridinium label used in combination with an automated (ACS 180, Bayer) analyzer for the detection of 3-PBA in urine resulted in a decrease in analysis time and a substantial increase in sensitivity. The automated system was able to analyze 100 samples in one hour with a 5-fold increase in assay sensitivity (IC50 of 0.30 ng/mL). Another advantage of this system is the saving of immunoreagents because of the lower concentration of antibodies required for binding reactions.

One approach for increased throughput in pesticide immunoanalysis is the development of lateral flow (dip stick) assays (61,62) or immunosensors combined with micro-channels (63) or flow-injection immunoaffinity analysis (64). A recent article by Seydack (65), presents a good overview on the development of biosensors based on nanoparticle labels and optical detection methods. Microfluidic lab-on-a-chip systems target biomarkers in physiological fluids with reduced sample, reagent, and assay time requirements, and therefore promise to have a significant impact on exposure analysis, especially in the field setting. The lab-on-chip systems are usually amenable to full automation and allow multiplexing of more than one target analyte but at the moment, they are still not used in practical applications.

A common trend in the human exposure analysis is a requirement for increased assay sensitivity to allow for detection of parent compounds or metabolites in the low parts per billion (ppb) range. One of the approaches for more sensitive assays has been the substitution of enzyme labels with more sensitive fluorescent probes (for a review see ref. 19). To overcome the inherent problem of high background fluorescence signals from biological material, probes that involve longer (far-red) wavelengths or longer fluorescence life-times seem to be most promising. Lanthanide chelates and lanthanide oxide nanoparticles have been successfully used as reporters in pesticide immunoanalysis (*14, 60, 66 67*). The advantages of europium oxide nanoparticles as fluorescent reporters include a large Stokes shift which decreases interference from scattered light, a sharp emission peak at the far-red (613 nm), a long-lifetime emission enabling time-gated detection and resistance to photobleaching.

Work by Ahn et al. (67) showed that the sensitivity of an immunoassay for the pyrethroid metabolite, 3-PBA, was increased by using europium oxide nanoparticle labels conjugated to the antigen. The assay was based on magnetic separation, and with an IC50 of 20 pg mL^{-1} showed about 1000-fold increase in sensitivity compared to the conventional 96-well plate assays with an enzyme label. Recently, Nichkova et al. (68) were able to demonstrate the use of biofunctionalized (IgG-PL-Eu:Gd$_2$O$_3$) nanoparticles as reporters in an indirect competitive fluorescence microimmunoassay for 3-PBA. Microarrays of BSA-PBA are fabricated by microcontact printing in line patterns (10 x 10µm) onto glass substrates. Confocal fluorescence imaging combined with internal standard (fluorescein) calibration was used for quantitative measurements. The non-optimized competitive microarray immunoassay had sensitivity in the low ppb range, which is similar to that of the conventional ELISA for 3-PBA. This work suggests the possibility for applying lanthanide oxide nanoparticles as fluorescent probes in microarray and biosensor technology, immunodiagnostics and high-throughput screening.

Other examples of promising high-throughput immunochemistry applications include chemiluminescent immunoassays based on microformat imaging using a charge-coupled device (CCD) camera (69). Ramanathan et al. (70) were able to use a portable module based on a photomultiplier tube (PMT) for the detection of pesticides in the field. More recently, Bhand et al. (71) used a novel immuno-array strategy for multicomponent analysis of two classes of pesticides (triazines and phenoxyalkanoic acids). The approach was based on cross-reactive arrays of specific antibody pairs coupled to chemometric pattern recognition. Undoubtedly, systems and devices first developed for monitoring pesticides and their metabolites in the environment (19, 72) will be well suited for human biomonitoring as well, making immunoassays even more valuable tools for human exposure studies.

References

1. Jakubowski, M.; Trzcinka-Ochocka, M. *J. Occup. Health* **2005**, *47*, 22-48.
2. Aprea, C.; Colosio, C.; Mammone, T.; Minoia, C.; Maroni, M. *J. Chromatogr. B* **2002**, *769*, 191-219.
3. Barr, D. B.; Needham, L. L., *J. Chromatog. B* **2002**, *778*, 5-29.
4. Harris, A. S.; Lucas, A. D.; Kramer, P. M.; Marco, M.-P.; Gee, S. J.; Hammock, B. D. In *New Frontiers in Agrochemical Immunoassay;* Kurtz, D. A.; Skerritt, J. H.; Stanker, L. H.,Eds; AOAC International: 1995, pp 217-235.
5. Krieger, R. I.; Ross, J. H.; Thongsinthusak, T. *Rev.Environ.Contam.Toxicol.* **1992**, *128*, 1-15.
6. Lavy, T. L.; Mattice, J. D. *ACS Symposium.Series 382* **1989**, 192-205.
7. Hoet, P. In *Biological monitoring of chemical exposure in the workplace;* World Health Organization: Geneva, 1996; Vol. 1, pp 1-19.
8. Katmeh, M. F.; Aherne, G. W.; Stevenson, D. *Analyst* **1996**, *121*, 1699-1703.
9. Önnerfjord, P.; Eremin, S.; Emneus, J.; Marko-Varga, G. *J. Chromatogr. A* **1998**, *800*, 219-230.
10. Denovan, L. A.; Lu, C.; Hines, C. J.; Fenske, R. A. *Int. Arch.Occup.Environ. Health* **2000**, *73*, 457-462.
11. Lu, C.; Anderson, L. C.; Morgan, M. S.; Fenske, R. A. *J. Toxicol.Environ. Health, A* **1998**, *53*, 283-292.
12. Lu, C. S.; Irish, R. M.; Fenske, R. A. *J. Toxicol. Environ. Health, A* **2003**, *66*, 2315-2325.
13. Divi, R. L.; Deland, A.; Fu, P. P.; Von Tungeln, L. S.; Schoket, B.;Eltz Camara, J.; Ghei, M.; Rothman, N.; Sinha, R.; Poirier, M.C. *Carcinogenesis* **2002**, *23*, 2043-2049.
14. Farmer, P. B.; Kaur, B.; Roach, J.; Levy, L.; Consonni, D.; Bertazzi, P.A.; Pesatori, A.; Fustinoni, S.; Buratti, M.; Bonzini, M.; Colombi, A.; Popov, T.; Cavallo, D.; Desideri, A.; Valerio, F.; Pala, M; Bolognesi, C.; Merlo, F. *Chemico-Biol. Interact.* **2005**, *153*, 97-102.
15. Jones, A. L.; Van Der Woord, M.; Bourrillon, F. *Ann. Occup. Hyg.* **2005**, *49*, 241-243.
16. Nichkova, M.; Marco, M.-P. *Env. Sci. Technol.* **2006**, in press.

154

17. Koivunen, M. E.; Gee, S. J.; Park, E.-K.; Lee, K.; Schenker, M. B.; Hammock, B. D. *Arch. Environ. Contam. Toxicol.* **2005**, 48, 184-190.
18. Shan, G.; Huang, H.; Stoutamire, D. W.; Gee, S. J.; Leng, G.; Hammock, B. D. *Chem.Res.Toxicol.* **2004**, 17, 218-225.
19. Schobel, U.; Barzen, C.; Gauglitz, G. *Fresenius J. Anal. Chem.* **2000**, 366, 646-658.
20. Biagini, R. E.; Murphy, D. M.; Sammons, D. L.; Smith, J. P.; Striley, C. A. F.; MacKenzie, B. A. *Bull. Environ. Contam. Toxicol.* **2002**, 68, 470-477.
21. Biagini, R. E.; Smith, J. A.; Sammons, D. L.; MacKenzie, B. A.; Striley, C. A. F.; Robertson, S. K.; Snawder, J. E. *Anal.Bioanal.Chem.* **2004**, 379, 368-374.
22. Lucas, A. D.; Jones, A. D.; Goodrow, M. H.; Saiz, S. G.; Blewett, C.; Seiber, J. N.; Hammock, B. D. *Chem. Res.Toxicol.* **1993**, 6, 107-116.
23. Jaeger, L. L.; Jones, A. D.; Hammock, B. D. *Chem. Res.Toxicol.* **1998**, 11, 342-352.
24. Striley, C. A. F.; Biagini, R. E.; Mastin, J. P.; MacKenzie, B. A.; Robertson, S. K. *Anal. Chim. Acta* **1999**, 399, 109-114.
25. Mastin, J. P.; Striley, C. A. F.; Biagini, R. E.; Hines, C. J.; Hull, R. D.; MacKenzie, B. A.; Robertson, S. K. *Anal.Chim.Acta* **1998**, 376, 119-124.
26. MacKenzie, B. A.; Striley, C. A. F.; Biagini, R. E.; Stettler, E.; Hines, C. J. *Bull. Environ. Contam.Toxicol.* **2000**, 65, 1-7.
27. Perry, M. J.; Christiani, D. C.; Mathew, J.; Degenhardt, D.; Tortorelli, J.; Strauss, J.; Sonzogni, W. C. *Toxicol.Ind. Health* **2000**, 16, 285-290.
28. Lyubimov, A. V.; Garry, V. F.; Carlson, R. E.; Barr, D. B.; Baker, S. E., *J. Lab. Clin. Med.* **2000**, 136, 116-124.
29. Rogers, K. R.; Van Emon, J. M. In *Immunoasssay for p-nitrophenol in urine;* ACS Symposium Series: 1994; pp 158-165.
30. Koivunen, M. E.; Dettmer, K.; Vermeulen, R.; Bakke, B.; Gee, S. J.; Hammock, B. D. *Anal. Chim. Acta* **2006**, submitted.
31. Lohse, C.; Jaeger, L. L.; Staimer, N.; Sanborn, J. R.; Jones, A. D.; Lango, J.; Gee, S. J.; Hammock, B. D. *J. Agric. Food Chem.* **2000**, 48, 5913-5923.
32. Shan, G.-M.; Wengatz, I.; Stoutamire, D. W.; Gee, S. J.; Hammock, B. D., *Chem.Res.Toxicol.* **1999**, 12, 1033-1041.
33. Nichkova, M.; Marco, M.-P. *Anal. Chim. Acta* **2005**, 533, 67-82.
34. Herve, J. J. In *The Pyrethroid Insecticides;* Leahey, J. P., Ed.; Taylor & Francis: London, United Kingdom, 1985; pp 343-425.
35. Class, T. J.; Kintrup, J. *Fresenius J. Anal. Chem.* **1991**, 340, 446-453.
36. Gianesi, L. P.; Silvers, C. S. *Florida Agricultural Statistic service. U.S. Department of Agriculture. July* **2001**.
37. Go, V.; Garey, J.; Wolff, M. S.; Pogo, B. G. T. *Environ. Health Persp.* **1999**, 107, 173-177.

38. EPA, List of Chemicals Evaluated for Carcinogenic Potential. *Washington, D.C.* **1994**.

39. Kolaczinski, J. H.; Curtis, C. F. *Food Chem.Toxicol.* **2004**, 42, 697-706.

40. Crawford, M. J.; Croucher, A.; Hutson, D. H. *Pest. Sci.***1981**, *12*, 399-411.

41. Miyamoto, J.; Kaneko, H.; Hutson, D. H.; Esser, H. O.; Gorbach, S.; Dorn, E., *Pesticide Metabolism: Extrapolation from Animals to Man.*; Blackwell Scientific Publications: Oxford, 1988; p 35.

42. Sipes, J. G.; Gandolft, A. J. In *Casarett and Doull's Toxicology: The basic Science of Poisons;* Amdur, M. O.; Doull, J.; Klaassen, C. D.,Eds.; Pergamon: New York, 1991; pp 64-98.

43. Ahn, K. C.; Watanabe, T.; Gee, S. J.; Hammock, B. D. *J. Agric. Food Chem.* **2004**, *52*, 4583-4594.

44. Matveeva, E. G.; Shan, G.; Kennedy, I. M.; Gee, S. J.; Stoutamire, D. W.; Hammock, B. D. *Anal. Chim. Acta* **2001**, *444*, (*1*), 103-117.

45. Angerer, J.; Ritter, A. *J. Chromatogr. B* **1997**, *695*, (2), 217-226.

46. Heudorf, U.; Angerer, J. *Environ. Health Persp.* **2001**, *109*, 213-217.

47. Ahn, K. C.; Ma, S.-J.; Tsai, H.-J.; Gee, S. J.; Hammock, B. D. *J. Anal. Bioanal. Chem.* **2006**, in press.

48. Summers, L. A., *The Bipyridinium Herbicides*; Academic Press Inc.: London, 1980; p 449.

49. Roberts, T. R.; Dyson, J. S.; Lane, C. G. *J. Agric. Food Chem.* **2002**, *50*, 3623-3631.

50. Van Emon, J. M.; Hammock, B. D.; Seiber, J. N. *Anal. Chem.* **1986**, *58*, 1866-1873.

51. Lee, X.-P.; Kumazawa, T.; Fujishiro, M.; Hasegawa, C.; Arinobu, T.; Seno, H.; Ishii, A.; Sato, K. *J. Mass Spectrom.* **2004**, *39*, 1147-1152.

52. Belluck, d. A.; Benjamin, S. L.; Dawson, T. *ACS Symposium Series* **1991**, *459*, 254-273.

53. Scribner, E. A.; Battaglin, W. A.; Goolsby, D. A.; Thurman, E. M. *Science Total Environ.* **2000**, *248*, 255-263.

54. Yoder, J.; Watson, M.; Benson, W. W. *Mutat. Res.* **1973**, *21*, 335-340.

55. Freeman, J. L.; Rayburn, A. L. *Environ. Toxicol.Chem.* **2005**, *24*, 1648-1653.

56. Ikonen, R.; Kangas, J.; Savolainen, H. *Toxicol. Lett.* **1988**, *44*, 109-112.

57. Catenacci, G.; Maroni, M.; Cottica, D.; Pozzoli, L. *Bull. Environ. Contam. Toxicol.* **1990**, *44*, 1-7.

58. Buchholz, B. S.; GFultz, E.; Haack, K. W.; Vogel, J. S.; Gilman, S. D.; Gee, S. J.; Hammock, B. D.; Hui, X.; Wester, R. C.; Maibach, H. I. *Anal. Chem.* **1999**, *71*, 3519-3525.

59. Baker, S. E.; Barr, D. B.; Driskell, W. J.; Beeson, M. D.; Needham, L. L., *J. Exp. Anal. Environ. Epidemiol.* **2000**, *10*, 789-798.

60. Gee, S. J.; Koivunen, M.E.; Nichkova, M.; Ahn, K. C.; Dosev, D.; Kennedy, I. M.; Hammock, B. D. *Abstracts of Papers*, 229th National Meeting of the American Chemical Society, San Diego, CA; American Chemical Society; Washington, DC, 2005; p.229.

61. Cuong, N. V.; Bachmann, T. T.; Schmid, R. D. *Fresenius J. Anal. Chem.* 1999, *364*, 584-589.

62. Weetal, H. W.; Rogers, K. G. *Anal. Lett.* 2002, *35*, 1341-1348.

63. Cummins, C. M.; Koivunen, M. E.; Stephanian, A.; Gee, S. J.; Hammock, B. D.; Kennedy, I. M. *Biosens. Bioelectron.* 2006, *21*, 1077-1085.

64. Krämer, P. M.; Franke, A.; Zherdev, A. V.; Yazynina, E. V.; Dzantiev, B. B. *Talanta* 2005, *65*, 324-330.

65. Seydack, M. *Biosens. Bioelectron.* 2005, *20*, 2454-2469.

66. Feng, J.; Shan, G. M.; Maquiera, A.; Koivunen, M. E.; Guo, B.; Hammock, B. D.; Kennedy, I. M. *Anal.Chem.* 2003, *75*, 5282-5286.

67. Ahn, K.; M.E., K.; Gee, S. J.; I.M., K.; Hammock, B. D. *Abstracts of Papers*, 227th National Meeting of the American Chemical Society, Anaheim, CA; American Chemical Society; Washington, DC, 2004; p.227.

68. Nichkova, M.; Dosev, D.; Gee, S. J.; Hammock, B. D.; Kennedy, I. M. *Anal. Chem.* 2005, *77*, 6864-6873.

69. Surugiu, I.; Svitel, J.; Ye, L.; Haupt, K.; Danielsson, B. *Anal. Chem.* 2001, *73*, 4388-4392.

70. Ramanathan, K.; Dzgoev, A.; Svitel, J.; Jonsson, B. R.; Danielsson, B., *Biosens. Bioelectron.* 2002, *17*, 283-288.

71. Bhand, S.; Surugiu, I.; Dzgoev, A.; Ramanathan, K.; Sundaram, P. V.; Danielsson, B. *Talanta* 2005, *65*, 331-336.

72. Rodriguez-Mozaz, S.; Lopez de Alda, M. J.; Marco, M.-P.; Barcelo, D., *Talanta* 2005, *65*, 291-297.

Chapter 11

Dosimetry and Biomonitoring following Golfer Exposure to Chlorpyrifos

Raymond A. Putnam and J. Marshall Clark

Massachusetts Pesticide Analysis Laboratory, Department of Veterinary and Animal Science, University of Massachusetts, Amherst, MA 01003

As golf and related recreational turfgrass uses become more urbanized, research on direct (immediate) and indirect pesticide exposure to humans, wildlife and adjacent environments becomes more and more essential. Human exposure risks to turfgrass pesticides can be correctly assessed only by knowing routes of exposure and the extent of absorbed dose. The present research emphasizes dosimetry (measuring pesticide residues on cotton suits, gloves, and air samplers worn by golfers) and biomonitoring (measuring pesticide metabolites in urine of golfers) in conjunction with environmental monitoring to determine recreational golfer exposure to chlorpyrifos. Resulting exposure estimates based on a 1h reentry interval following full-course and full-rate applications of chlorpyrifos were significantly less than established USEPA OPP reference dose (Rfd) criteria. These low exposures were successfully mitigated using a partial-course application strategy.

Human exposure following the application of pesticides for the proper management of turf environments continues to be a concern. This concern is germane given the level and frequency of pesticide use, the extent of human activities and time spent on turfgrass, and the exposure potential for infants, children, and adults alike. Considerable effort has been expended in the determination of applicator exposure issues and the means to mitigate problematic exposure situations during the mixing and application of pesticides. There are other potential exposure concerns, however, for all who reenter turfgrass environments following pesticide applications (1) and little research exists on direct assessments of golfer exposure to pesticides applied to turfgrass (2,3). Golfers elicit unique behaviors in this recreational setting not usually duplicated with pesticide applicators or agricultural workers. Additionally, golf courses are usually open everyday of the week, leaving little time between pesticide application and reentry. A third prominent factor is the uniqueness of the turfgrass system itself compared to other agricultural settings (2). Thus, a comprehensive evaluation of the exact exposure that a golfer receives while playing golf and the health implication, if any, of that exposure is necessary. Proper safeguards can then be developed to eliminate or reduce future exposures.

The primary route of exposure from pesticide-treated turf involves dermal uptake from dislodgeable foliar pesticide residues (DFRs, pesticide residues present on the treated foliage available by contact or abrasion for skin absorption) (2). It is expected that a larger proportion of the applied pesticide will remain on the turfgrass leaves because of the dense canopy inherent in turfgrasses compared to agricultural cropping situations, where a substantial proportion of the pesticide reaches the soil surface directly (4,5). Thus, dermal exposure to DFRs on turfgrass is expected to be significant. Nevertheless, most turfgrass cultivars used for lawns, golf courses, etc., have substantial waxy layers associated with the external surfaces of their blades and all grasses produce organically-rich thatch/mat layers. These aspects of the turfgrass plant are expected to compete with transfer of pesticides to exposed hands, legs, etc. and reduce dermal exposure levels.

The next most significant exposure route involves inhalation of airborne pesticide residues (volatile pesticides and residues associated particulate, such as aerosols and dust particles) by the lung during breathing. Although usually not considered as significant as dermal exposure, the respiratory route is toxicologically relevant due to its high rate of absorption and direct interaction with the circulatory system, allowing the rapid and extensive distribution of airborne pesticides through the body.

The oral route of exposure via the gastrointestinal tract is considered the least extensive and occurs primarily via hand to mouth contact, a situation more relevant for children rather than adults. Preliminary evidence has indicated that golf balls, tees, etc., do not acquire large amounts of pesticides and are not efficient means to transfer significant levels of pesticides to golfers (6).

This current study emphasizes dosimetry (measuring pesticide residue on full body cotton suits, gloves, veils, and personal air samplers) and biomonitoring (measuring pesticide metabolites in collected urine). Dosimetry together with concurrently collected dislodgeable foliar and airborne residue data, provides the basis for modeling exactly how much pesticide is transferred from the turfgrass to an individual during a round of golf. Biomonitoring is increasingly being used to quantify human exposure to pesticides because it requires fewer assumptions (7,8), such as exposure routes or rates of transfer, clothing penetration and/or dermal absorption rates. The advantage of biomonitoring is that it measures the amount of pesticide in collected human tissues and fluids, which is directly related to absorbed dose. Thus, when the pharmacokinetics (absorption, distribution, metabolism, elimination) of a compound is known, and when a suitable biomarker is available (major urinary metabolite), biomonitoring presents the most complete picture for assessing whole body absorbed dose and the health implications of that exposure (9,10). Nevertheless, the biomonitoring of human subjects may not always be possible due to questions of safety. In many situations, it may also be impractical to test a variety of exposure situations and mitigation strategies utilizing human subjects because of time and funding constraints. Additionally, not all pesticides produce major urinary metabolites, and few pesticides have human pharmacokinetic data available. Fortunately, once the relationship between absorbed dose and chemical disposition is known, accurate exposure models can be developed to predict human exposure without the need for further human subjects.

Biomonitoring data alone will not provide information about the source, magnitude, or frequency of exposure. One reason is that most contemporary pesticides have short half-lives in the human body and acute exposures could be underestimated. The same levels could be measured in urine, for example, as the result of a single exposure event or multiple smaller exposures. For this reason, biomonitoring is often used in conjunction with complimentary environmental data and modeling approaches to estimate exposure (9,10). In this study, we estimated golfer exposure to chlorpyrifos utilizing three independent techniques. The direct and simultaneous determination of dosimetry and biomonitoring data, along with concurrently collect environmental residues (dislodgeable foliar and airborne) provides a novel and complete information base on how much pesticide is actually transferred to and absorbed by a golfer playing golf.

Methodology

Chlorpyrifos

Chlorpyrifos (CHP, *O,O*-diethyl *O*-3,5,6-trichloro-2-pyridyl phos-phorothioate, Fig. 1), a well studied phosphorothioate insecticide of moderate mammalian toxicity, was selected for this study since its metabolite, TCP (3,5,6-trichloro-2-pyridinol), is stable, readily cleared in urine after dermal and oral doses ($t_{1/2}$ = 27 h), and human toxicokinetic studies are available (*11*). The US EPA Registration Eligibility Document (*12*) lists a dermal absorption value of 3 %, however a dermal absorption rate of 9.6 % per 24 hr was assigned by Kreiger et al., (*10*) and Thongsinthusak (*13*). CHP has low water solubility (1.39 mg/L) and a moderate vapor pressure (2 x 10^{-5} mm Hg @ 25°C). The current US EPA OPP chronic reference dose (Rfd; lifetime daily exposure without adverse health effect) for chlorpyrifos is 3.0 µg/Kg/d (*14*).

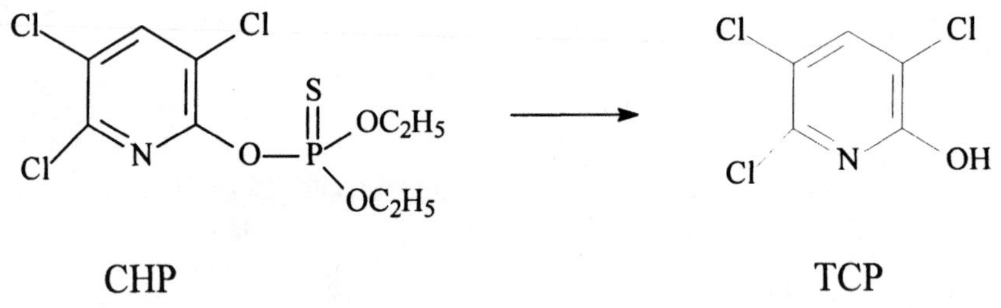

Figure 1. Chlorpyrifos and its principal urinary metabolite, TCP.

Turfgrass Plots, Application of Pesticides, and Collection of Environmental Pesticide Residues

Turfgrass Plots

All experiments were conducted at the University of Massachusetts Turfgrass Research Center in South Deerfield, MA. Two circular (10 m radius) turfgrass plots with established "Penncross" creeping bentgrass were used for the collection of environmental pesticide residues (airborne and DFRs) as previously described (*5,15*). Additionally, a 100 m X 20 m rectangular bentgrass turfgrass plot was used for the concurrent collection of dosimetry and biomonitoring data. All plots were maintained as golf course fairways (mowed at a height of 1.3 cm three times per week and irrigated as needed to prevent drought stress).

Pesticide Applications

A Rogers Sprayer (35-40 psi), equipped with a wind foil, skirt, and twelve spray nozzles fitted with VisiFlo® Flat Spray Tips, was used for all applications. Dursban Pro® (23.5% chlorpyrifos) was applied at the maximum labeled US EPA approved rate prior to 2002 of 4 lbs a.i./acre. 1 gallon (4 lbs a.i./acre) of formulated product was mixed into 50 gallons of water and applied at approximately 100 gallons/acre. All applications were immediately followed with 1.3 cm of post-application irrigation.

Airborne and Dislodgeable Foliar Residues

These experiments were carried out on the paired circular bentgrass plots. Airborne residues of chlorpyrifos ($\mu g \ m^{-3}$) were determined with a single TF1A high-volume air sampler located in the center of each circular plot using the methodology of Kilgore et al., (*16*) as modified by Murphy et al. (*4*).

Dislodgeable foliar residues (DFRs) were determined using a water-dampened cheesecloth wipe (*5*) and with the California roller device (CA roller) (*17*). Cloth wipe samples were collected in triplicate at each plot at 0.25, 1, 2, and 5 h post-application. CA roller samples were collected in triplicate at each plot at 1, 2, and 5 h post-application.

Collection and Analysis of Dosimetry and Biomonitoring Samples

At the same time that airborne and dislodgeable foliar pesticide residues were being collected, exposure to researchers simulating the play of golf was determined by dosimetry and biomonitoring studies using the rectangular bentgrass plots. Each experiment utilized eight volunteers (one foursome for dosimetry, a second for biomonitoring) simulating the play of an 18-hole round of golf over a period of 4 h. Volunteers for the dosimetry and biomonitoring groups were from the UMASS Environmental Toxicology and Risk Assessment Program (School of Public Health) and the Department of Veterinary and Animal Science. A protocol that described the research and protected the rights of the volunteers has been approved by the Human Subjects Review Committee, UMASS. The approved protocol, including an informed consent form, was reviewed with participants prior to their participation.

In the "standardized" 18-hole round of golf, each player walked 6,500 yards, hit a ball 85 times, and took 85 practice swings. Clubs were rotated in an appropriate way, balls teed-up, divots replaced, two putts taken each hole, and clubs wiped clean between shots using a golf bag towel. Each 4-h "round" of golf began 1 h following the completion of post-application irrigation. During experiments that simulated pesticide treatment of tees and greens only, golfers spent 3 min on the treated tee-boxes, 7 min on the untreated surface with continuous walking, and 3 min putting-out on the treated greens for each hole.

Whole Body Dosimetry (passive and active dosimetry)

Participants in the dosimetry group wore a long-sleeved shirt and long pants made from a single layer of white, sanforized 100 % cotton (Universal Overall Corp, Chicago Il), two pairs of cotton gloves (VWR Scientific), and veil (7.5 x 14' 200-thread count cotton fabric) attached with safety pins to the back of hat. Participants changed to a fresh pair of double gloves at the two-hour mark. This cotton clothing served as a passive collection medium for DFRs from treated turfgrass (*10*). It was removed at the end of the golf round and sectioned as follows for analysis: lower arms, upper arms, torso, lower legs, upper legs/waist, gloves, and veil.

Residues of chlorpyrifos were extracted from dosimeters and DFR cloths using hexane and after a liquid-liquid partition cleanup, analyzed by gas chromatograph (GC) equipped with a nitrogen-phosphorous detector (Agilent Technologies 6890, Agilent Technologies, Inc, Wilmington, DE) using a fused silica column (DB-5 liquid phase, 30M x 0.25 mm i.d., 0.25 μm film thickness, J & W Scientific). Daily performance amendments were made for all cloth samples and recoveries from fortified cloth samples ranged from 77- 115%.

Inhalation exposure was measured using a personal air sampling pump (AirChek 52, SKC, Eighty Four, PA) calibrated to a flow of 2.0 liter of air per minute fitted with a glass fiber OVS sampling tube (140/270 mg XAD-2, SKC) attached to the volunteers' collar (*18,19*). Following sampling, the front sorbent section and the glass microfiber were combined into an 8 ml vial and the rear sorbent bed into another. Pesticide residues were desorbed for 1 h with 2.0 ml toluene containing 2 μg/ml triphenyl phosphate (TPP) as an internal standard. Extracts were analyzed for chlorpyrifos using a GC equipped with a mass selective detector (Hewlett-Packard 5971) operated in selected ion-monitoring mode (SIM). The amounts of chlorpyrifos detected in the air samplers were adjusted to the adult breathing rate during light activity (21 L /min). Recoveries for personal air sampler tubes fortified between 0.3 and 2.0 μg of chlorpyrifos was 94.7 ± 4.7 %.

Urinary Biomonitoring

The biomonitoring group wore short sleeve shirts, shorts, golf caps, ankle socks and golf shoes. To estimate the total absorbed dose following chlorpyrifos exposure, urinary biomonitoring was conducted for TCP (*9,20*). Recoveries of distilled water and pooled control urine fortified with the TCP over a range of 2 – 100 µg/l was 101.7 ± 11.2 %, N= 38.

Urine samples were collected and analyzed for TCP the day before exposure (27 h), and then for a 27 h interval following chlorpyrifos exposure (estimated time to excrete ½ of the total TCP). To determine whole body dose of chlorpyrifos, the amount of TCP was determined and divided by the urinary excretion factor of 0.4. This factor represents the ratio of molecular weights of TCP (198) and CHP (350.6) (i.e., 198/350.6 = 0.56) and the fraction of the absorbed dose expected to be excreted in urine (0.72). The fraction expected to be excreted in urine is based on a human study in which an average of 72% of orally administered CHP was excreted in the urine as TCP (*11*). An additional correction factor of 2 was added because dose was estimated from half-life (*t* = ½) excretion. Volunteers were instructed to avoid exposure to any chlorpyrifos during the week prior to the golf-related exposure. Because the half-life is approximately 27 h, the volunteers would have reached a steady state TCP elimination on the day before the golf-related exposure (*9*).

Hazard Assessment of Golfer Exposure

Golfer hazard was assessed using the US EPA Hazard Quotient calculation (*2*) independently utilizing exposure estimates from dosimetry versus biomonitoring data. The estimated absorbed dose (AD) was divided by the chronic US EPA chronic reference dose (Rfd) to give a Hazard Quotient (AD/Rfd =HQ). A HQ value less than or equal to 1.0 indicates that the residues present are at concentrations below those that are expected to cause adverse effects to humans. A HQ value greater than 1.0 does not necessary infer that adverse effect will occur, but rather that the absence of adverse effects is less certain.

Results

Determination of Exposure by Dosimetry

Total chlorpyrifos residues collected on whole body dosimeters and onto individual personal air samplers are summarized in Figure 2. Whole body dosimeters collected an average of 305 ± 58 μg following full course applications. The absorbed dermal dose (ADD) for a 70 Kg adult using a 9.6 % dermal penetration factor was calculated as 0.42 μg chlorpyrifos/Kg. The absorbed inhaled dose (AID) assuming a breathing rate of 21 L/min and an absorption rate of 100% was calculated as 0.18 μg/Kg. Application of chlorpyrifos to only the tees and greens resulted in a 76-81% reduction in whole body dosimetry and a corresponding 75-84% reduction in airborne residues (Fig. 2).

Figure 3 illustrates the distribution of pesticide collected onto various body regions by the dosimeter groups. One of the most pronounced findings from the dosimetry data was that the major route of potential exposure to golfers was dermal where it accounted for > 92% of all transferable residues. Additionally, it was previously thought that the hands were the primary route for dermal exposure *(15,21)*. However, we found the major route for exposure is the lower legs. The lower leg consistently was the most highly contaminated collector, followed by pants (upper leg to waist) and torso. When combined with the residues on hands and lower arms (forearms), the areas generally exposed on most golfers, approximately 85% of the total pesticide residues transferred to the whole body dosimeters are present in these areas.

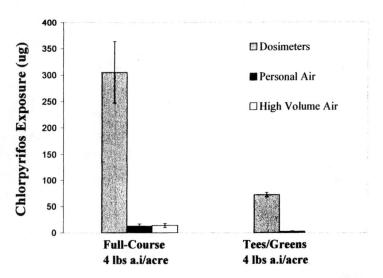

Figure 2. Total chlorpyrifos residues (exposure) collected on whole body dosimeters and onto individual personal and high volume air samplers.

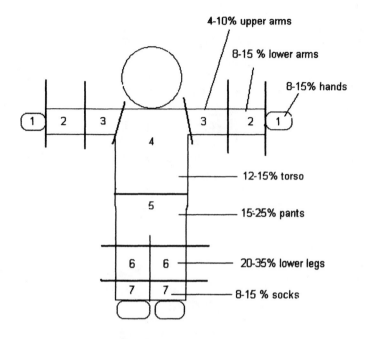

Figure 3. Distribution of chlorpyrifos residues collected on dosimeters.

Hazard Quotient (HQ) Determination based on Dosimetry

A HQ based on dosimetry was calculated by dividing the combined dermal and inhalation doses by the USEPA Rfd.

$$HQ = \frac{0.42 \ \mu g/Kg \ (dermal) \ + \ 0.18 \ \mu g/Kg \ (inhalation)}{3 \ \mu g/Kg/d \ (EPA \ OPP \ chlorpyrifos \ Rfd)}$$

$$HQ = 0.2$$

Determination of Exposure by Biomonitoring.

Table 1 summarizes the biomonitoring data and estimates the actual whole body dose of absorbed chlorpyrifos. For all volunteers, baseline (pre-application exposure) urinary TCP levels (average of 5.14 µg/L TCP; equivalent to 0.245 µg CHP/Kg ± 0.111, N = 24) were consistent with reference concentration levels for the general population as reported by Hill et al., (22) and background levels reported by Byrne et al., (9). Baseline urinary measurements were subtracted from the turfgrass exposure-related TCP concentrations, so that reported exposure estimates represent only the absorbed dose resulting from exposure during simulated golf activities (Table 1). The mean whole body dose of chlorpyrifos determined by biomonitoring at the full rate and full-course application scenario was 1.06 µg/Kg/d. Dividing this value by the US EPA Rfd for chlorpyrifos (3 µg/Kg/d) yields a HQ value of 0.35.

Substantial reductions in whole body doses of chlorpyrifos were apparent when applications were restricted to tees and greens, and were consistent with the reductions seen for dosimetry samples during similar application scenarios. There was an 87% overall reduction in chlorpyrifos exposure following applications to only tees and greens versus whole course applications, resulting in a HQ of 0.047.

Table 1. Total absorbed doses and HQ values determined from biomonitoring following full- and partial-course applications of chlorpyrifos.

Application Scenario	Absorbed Dose (µg/Kg)[1]	HQ
full-course, 1 hr re-entry	1.058 ± 0.409	0.352
tees & greens, 1 hr re-entry	0.140 ± 0.069	0.047

[1] Post exposure chlorpyrifos equivalents - pre-sample chlorpyrifos equivalents (day^{-1}).

Monitoring Environmental Pesticide Residues.

Dislodgeable Foliar Residues (DFRs)

As evidenced by chlorpyrifos DFRs determined by cloth wipe samples, a dramatic decline occurred during the first hour drying-in period following post-application irrigation, followed by a slower but steady decline over the next 4 h (Fig. 4). Cloth wipe estimates of chlorpyrifos declined from 0.09 µg/cm^2 at 0.25 h to 0.04 µg/cm^2 at 1 h (55% reduction), and then to 0.01 µg/cm^2 over the next 4 h (an additional 49% reduction). Averaging DFRs determined from cloth wipe measurements over the 4 h period when golfer exposure occurred (1-5 h post-application and irrigation) resulted in a mean DFR value (± S.E.) of 0.0249 ± 0.013 µg chlorpyrifos/cm^2.

DFRs determined from CA roller samples elicited a similar dissipation pattern compared to that obtained with cloth wipe samples except that the levels of DFRs are reduced ~50 % at all time intervals examined (Fig. 4). With this technique, a mean 4 h DFR value (± S.E.) of 0.012 ± 0.0046 (N=12) μg chlorpyrifos/cm^2 was determined. The CA roller technique was selected over the cloth wipe technique for use in calculating a dermal transfer factor because it provided a standardized, reproducible process necessary to estimate the amount of DFRs available for transfer from treated turf (*17,23*). The dermal chlorpyrifos transfer factor (TF) relevant to golfer activities was calculated using the method of Zweig et al. (*24*).

Dermal Exposure (μg) =TF (cm^2/hr) x DFR (ug/cm^2) x 4 h

Dermal exposure (μg) was derived from whole body dosimeters and DFR was the mean 4 h DFRs obtained by CA roller. The calculated dermal transfer factor for chlorpyrifos was 6363 ± 583 cm^2/h.

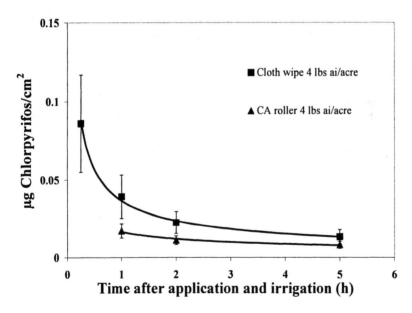

Figure 4. Dislodgeable foliar residues (DFR) of chlorpyrifos over the first 5 h following post-application irrigation.

Airborne Pesticide Residues

Using the measured air concentration of chlorpyrifos determined by high-volume air sampling, an average inhaled dose (D_i) for a 70 Kg adult playing an 18-hole round of golf was estimated using the same procedure as described for the personal air samplers:

$$C \times R \times 4 \text{ h}/70\text{Kg} = D_i$$

where C = average concentration of pesticides in air determined by high-volume air sampling, R = adult breathing rate during moderate activity (21 L /min), and D_i = inhaled dose of pesticides. An average of 13.5 µg ± 3.7 chlorpyrifos was estimated to be absorbed via the lung at this moderate breathing rate, a value similar to that obtained using the personal air sampling technique worn by dosimetry volunteers. The estimates of the IHQ using high volume air samplers (0.064) and the personal air samplers (0.060) were in good agreement.

Discussion

In the current study, exposure estimates based on a 1 h reentry interval following full-course applications of chlorpyrifos were substantially below current US EPA Rfd values. Following full course chlorpyrifos application at 4 lbs a.i./acre, total golfer exposure as measured by dosimetry resulted in a HQ of 0.2. Using the mean whole body dose of chlorpyrifos from urinary biomonitoring, a HQ value of 0.35 was obtained. These values are in good agreement and are both significantly below 1, indicating a wide margin of safety. As expected, exposures following full course applications were successfully attenuated using a partial course application strategy. Overall, there was an ~ 87% reduction in chlorpyrifos exposure following applications to only tees and greens versus whole course applications as measured by both biomonitoring and dosimetry.

Chlorpyrifos is a high risk insecticide that has both high volatility and inherent high toxicity (relatively low Rfd). Even with these characteristics, its potential for exposure that would result in hazardous human health implications following the play of golf is not likely. Newer pesticides that do not share the potentially harmful chemistry evident with chlorpyrifos and which are applied at lower rates are expected to pose an even lower risk when evaluated by biomonitoring and dosimetry approaches. Also, it is also unlikely that golfers will encounter these worst case exposures on every round of golf over a period of many years.

Exposure estimates based on dosimetry were calculated by combining the residues collected on the entire whole body suit and multiplying by the 9.6%

dermal penetration rate, and adding estimated inhaled dose based on a breathing rate indicative of light activity. Overall chlorpyrifos exposure estimated from biomonitoring (1.058 µg/Kg, HQ = 0.35) was 1.8 x higher than those estimated from dosimetry (0.599 µg/Kg, HQ = 0.2) and suggested that either the assumptions used for assessing inhaled dose were conservative (i.e., breathing rate for golfers was higher than thought) or the assumptions used for estimating dermal dose were conservative (i.e., the dermal penetration rate was higher than modeled). To estimate dermal dose, we used the higher published dermal absorption rate of 9.6 % (*13*). Nonetheless, dermal absorption is known to be increased due to occlusion (*25*), use of sunscreens (*26*), skin moisture (*27*), etc. In addition, the extent of absorption though skin in also influenced by body location (*28*) as well as concentration of material, presence of carrier or formulation, and temperature, making estimates of dermal absorption during real-life scenarios problematic.

Nevertheless, the good agreement amongst these different techniques indicate that the biomonitoring and dosimetry techniques, combined with the measurement of environmental residues, provides a thorough picture of transferable pesticide residues and golfer exposure, and forms the basis for predicting absorbed dermal dose (ADD) to golfers solely by using the standardized CA roller procedure:

$$ADD = S \times P \times 4hr / 70 \text{ Kg}$$

where S is determined by multiplying the mean 4 h DFR value determined from CA roller technique (0.012 ± 0.0046 µg chlorpyrifos/cm^2) by the empirically determined dermal transfer coefficient of 6363 ± 583 cm^2/hr, and by the dermal permeability (P = 9.6%).

Dermal pesticide exposure has been found to be the most significant route of potential exposure to golfers (> 92% of total residues) and is thought to occur primarily by the transfer of DFRs to an individual's skin and/or clothing. The lower legs, hands and lower arms are the most vulnerable routes of exposure. However, when considering the relative absorption rates of chlorpyrifos via dermal (9.6%) versus inhalation (100%) penetration, inhalation exposure may account for 30% of the overall absorbed dose. Nevertheless, the relative contribution of potential inhalation exposure for those pesticides that are more water soluble and less volatile than chlorpyrifos is expected to be far less.

DFRs rapidly declined over the first hour "drying-in" period and the potential for dermal exposure is dramatically reduced following a 1 h post-application and irrigation interval. We have previously reported that DFRs are reduced by approximately 80% by post-application irrigation (*5*), and our new findings show that DFRs are reduced by yet another ~50% simply by enforcing a

1 h reentry interval. These findings are again encouraging and indicate that future studies of operational practices to attenuate exposure (e.g. reentry intervals, irrigation, application strategies, alternative chemicals and IPM strategies) are highly likely to be effective in Best Management Strategies for pesticide use on turfgrass.

Acknowledgements

The authors express their appreciation to the USGA, USDA, and the New England Regional Turfgrass Foundation for partial funding of this work, and to Dow AgroSciences and Bayer Corporation for materials and technical assistance, and to Jeff J. Doherty, Robin Edwards and the volunteer golfers for helping complete this work.

References

1. Sigler WV, T., C.P., Throssel, C.S., Bischoff, M., Turco, R.F. *Environmental fate of fungicides in the turfgrass environment: A minireview;* Am. Chem. Soc.; Washington, D.C., 2000.
2. Clark, J. M., Kenna, M.P. *Handbook of Pesticide Toxicology;* Kreiger, R., Ed., Academic press; San Deigo, CA, 2001; Vol1, pp 203-242.
3. Bernard, C. E.; Nuygen, H.; Truong, D.; Krieger, R. I. *Arch. Environ. Contam. Toxicol.* **2001**, *41*, 237-240.
4. Murphy, K. C., R.J. Cooper and Clark, J.M.; *World Scientific Congress of Golf II*; Cochran, A.J., Farrally, M.R., Eds.;. E. & F.N. Spon, London, 1994, pp. 505-510.
5. Clark, J. M., Roy, G., Doherty, J.J., Cooper, R.J. *Fate and Management of Turfgrass Chemicals*; J. M. Clark, and Kenna, M.K., Ed.; ACS Symposium Series 743; Am. Chem. Soc., Washington DC, 2000. pp 294-313.
6. Cisar, JL, S., R.H., Sartain, J.B., Borget, C.J.; *USGA Turfgrass and Environmental Research Online 2002*, **2002**, *1*.
7. USEPA; *Occupational and Residential Exposure Test Guidelines: OPPTS 875.2600- Biological Monitoring. EPA 7122-C-96-271*; US EPA, Office of Prevention, Pesticides and Toxic Substances, 1996.
8. Franklin, C. A., Fenske, R.A., Greenhalgh, R., Mathieu, L., Denley, H.V., Leffingwell, J.T., Spear, R.C.; *J. Toxicol. Environ. Health.* **1981**, *7*, 715-731.
9. Wollen, B., Marsh, J., Laird, W., Lesser, J.; *Xenobiotica.* **1992**, 22, 983-991.

10. Byrne, S. L.; Shurdut, B. A.; Saunders, D. G.; *Environ. Health Perspect.* **1998**, *106*, 725-731.

11 Krieger, R. I., Bernard, C. E., Dinoff, T. M., Fell, L., Osimitz, T. G., Ross, J. H., Ongsinthusak, T. *J. Expo. Anal. Environ. Epidemiol.* **2000**, *10*, 50-57.

12. Nolan, R. J., Rick, D. L., Freshour, N. L., *Toxicol. Appl. Pharmacol.* **1984**, *73*, 8-15.

13 USEPA; *Interim Reregistration Eligibility Decision for Chlorpyrifos*; Office of Prevention, Pesticides and Toxic Substances, 2002.

14. Thongsinthusak, T. *Determination of Dermal Absorption of Chlorpyrifos in Humans;* California Environmental Protection Agency, Department of Pesticide Regulation, 1991.

15. *The Pesticide Manual*; Tomlin, C. D. S., 13 ed.; British Crop Protection Council: Hampshire, UK, 2003, pp 173-174.

16. Murphy, K. C., Cooper, R. J., Clark, J. M.; *Crop Sci.* **1996**, *36*, 1446-1454.

17. Kilgore, W., Fischer, C., Rivers, J., Akesson, N., Wicks, J., Winters, W., Winterlin, W.; *Residue Rev* **1984**, *91*, 71-101.

18. Fuller, R., Klonne, D., Rosenheck, L., Eberhart, D., Worgan, J., Ross, J.; *Bull Environ Contam Toxicol* **2001**, *67*, 787-794.

19. OSHA; *Method 62: Chlorpyrifos, DDVP, Diazinon, Malathion and Parathion in Air.* US Dept. of Labor, Occupational Safety & Health Administration, Organic Method Evaluation Branch, Occupational Safety and Health Administration, Analytical Laboratory, 1986.

20. OSHA; *Method 63:Carbaryl (Sevin)*; US Dept. of Labor, Occupational Safety & Health Administration, Organic Method Evaluation Branch, Occupational Safety and Health Administration, Analytical Laboratory, 1987.

21. Olberding, E. L.; *Determination of Residues of TCP in Urine by Capillary Gas Chromatography with Mass Selective Detection*; Dow AgroSciences, Global Environmental Chemistry Laboratory, 1998.

22. Murphy, K. C., Cooper, J.R., Clark, J. M.; *Crop Sci.* **1996**, *36*, 1455-1461.

23. Hill, R. H., Jr.; Head, S. L.; Baker, S.; Gregg, M.; Shealy, D. B.; Bailey, S. L.; Williams, C. C.; Sampson, E. J.; Needham, L. L.; *Environ. Res.* **1995**, *71*, 99-108.

24. Zweig, G.; Leffingwell, J. T.; Popendorf, W.; *J Environ Sci Health B* **1985**, *20*, 27-59.

25. Poet, T. S.; *Toxicol Sci* **2000**, *58*, 1-2.

26. Brand, R. M., Spalding, M., Mueller, C.; *J Toxicol Clin. Toxiol.* **2002**, *40*, 827-832.

27. Meuling, W. J., Franssen, A. C., Brouwer, D. H., van Hemmen, J. J.; *Sci Total Environ* **1997**, *199*, 165-172.

28. Wester, R. C., Maibach, H. I., Bucks, D. A., Aufrere, M. B.; *J Toxicol Environ Health* **1984**, *14*, 759-762.

Chapter 12

Assessment of Pesticide Exposures for Epidemiologic Research: Measurement Error and Bias

Shelley A. Harris

Department of Epidemiology and Community Health and Center for Environmental Studies, Virginia Commonwealth University, P.O. Box 843050, Richmond, VA 23284–3050 (saharris@vcu.edu)

Although numerous epidemiologic studies have been conducted to evaluate acute and chronic health effects associated with pesticide exposures, results of these studies are not consistent, may often be biased, and are generally not supported with accurate pesticide exposure data. Inadequate measurement of pesticide exposure, or preferably dose, is a major factor limiting the value of study results. Since it is generally not possible to measure pesticide exposures retrospectively, and not cost-effective or practical to measure exposures prospectively, alternative techniques must be developed and evaluated for use in epidemiologic research. Past exposure assessment methods, their associated biases, and current efforts are described.

Introduction

Of the many potentially hazardous occupational and environmental exposures we experience in daily life, very few cause the anxiety and concern that are commonly associated with pesticide exposure. Possible residues of pesticides in foods, especially those consumed in large quantities by children, have fuelled the organic movement and have resulted in considerable pressure on governments to re-evaluate and modify the risk assessment process. Unintentional exposure of bystanders, either real or perceived, and subsequent complaints of acute symptoms, environmental sensitivity or multiple chemical sensitivity, coupled with concern over harm to children have resulted in the posting of almost all professional pesticide applications and the creation of spraying pre-notification registries. Despite numerous epidemiologic studies in both adults and children, the health risks associated with chronic occupational or environmental exposures to commonly used pesticides are largely unknown and strongly debated. The measurement of pesticide exposure is one of the key issues that limits the value of these studies and contributes to the controversy that surrounds the results.

The exposure, disease, and confounder triangle is well known to the students and practitioners of epidemiology, and much time is allocated to the assessment of the accuracy of disease or outcome measurement and the evaluation or control of confounding. Valid measurement of exposure, although recognized as an important issue, has in the past, not been given adequate consideration in the design, conduct, analysis or interpretation of epidemiologic research [1]. More recently in occupational and environmental studies of disease, exposure measurement has received considerably more attention as we look to detect increasingly lower chronic health risks associated with chemical and physical exposures.

The herbicide 2,4-dichlorophenoxyacetic acid (2,4-D) is one pesticide that fits the profile of a compound that: has been extensively studied in both laboratory animals and humans; has been associated with controversial health risks in humans; has caused considerable public concern, and; has not had exposures measured adequately in epidemiologic studies of disease [2,3]. Although a number of adverse health effects have been reportedly associated with 2,4-D and related herbicides, and a number of effects, such as endocrine disruption, have been hypothesized, the majority of toxicologic and epidemiologic work has been conducted to evaluate the association between exposure to 2,4-D and cancer. Therefore, current methods used to assess chronic pesticide exposures and issues related to exposure measurement error will be discussed. Finally, the development of exposure and dose prediction models will be introduced.

Issues in Epidemiologic Studies of Pesticide Exposures

One of the greatest barriers to obtaining useful results in epidemiologic studies is the lack of adequate exposure measures or more specifically, the lack of dose data. Inaccurate assessment of exposure can seriously bias risk estimates and can result in the dilution of risk estimates or the erroneous identification of statistical associations that truly do not exist. Many of the studies evaluating the health risks associated with exposure to 2,4-D have been hindered by poor design, poor assessment of exposure and confounding variables, and mixed exposures (other phenoxy herbicides and their associated dioxins). Results of these studies have been conflicting and it is not possible to determine whether this is due to poor assessment of exposure, the lack of control of confounding, recall bias, or the absence of a true carcinogenic effect. Monson has pointed to the need for more specific assessment and measurement of exposure to 2,4-D for future case-control and cohort studies [4]. The concept of exposure in epidemiology, by necessity, takes on a very different meaning when compared with its interpretation in toxicology and risk assessment.

Exposure and Dose Distinguished

The terms exposure and dose are often used interchangeably, generally with the understanding that important distinctions exist between the two. In occupational settings, exposure refers to the concentration of an agent at the boundary between an individual and the environment as well as the duration of contact between the two [5]. Dose, or more specifically, internal dose, is the amount of an agent that is absorbed, inhaled, or ingested into the body and is generally expressed over a given time. The biologically effective dose is the amount of the active parent compound or metabolite that reaches a target organ and exerts an early biological effect that may eventually lead to clinical disease (see Figure 1).

A number of host factors which may be both genetic and environmental will affect the relationship between potential exposure and the resulting biologically-effective dose. Although all individuals working within a defined work area may have the same opportunity for potential exposure, activity patterns and location of

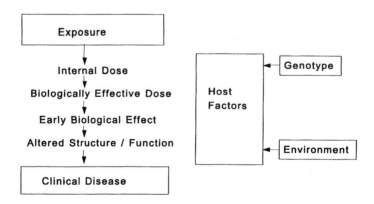

Figure 1. The path of exposure to clinical disease

work areas relative to primary sources of exposure will affect these levels. For agents in which inhalation is the primary route of exposure, host factors such as breathing rates and the amount of physical exertion can modify the resulting internal dose. When the primary route of exposure is dermal, the internal dose can be modified by factors such as the area and location of skin exposed, skin damage, the number of hair follicles, and other environmental considerations such as the temperature and humidity, and the presence of other compounds on the skin [6-8]. The relationship between potential exposure and internal dose will also be modified by work-hygienic practices such as wearing gloves or using respirators, and may be further modified by attitudes, avoidance behaviours, and risk perceptions, to name a few. Individual differences in the metabolism, detoxification or activation of a compound will alter the biologically effective dose and differences in susceptibility will influence the early biological effect. Clearly, the path between potential exposure in an occupational or environmental setting and the resulting biologically effective dose is complex.

In occupational settings, inhalation exposure may be estimated with the use of air monitoring of work areas, or personal monitoring of the breathing zones. Dermal exposures may be evaluated with the use of patches, whole body dosimetry, or fluorescent tracers. Depending on the physical characteristics of the agent of interest and the primary routes of exposure, a combination of these methods may be used to evaluate exposure. To convert these exposure estimates to internal dose estimates, constant breathing rates, body weights, dermal absorption rates, and a direct linear relationship between exposure and dose are generally assumed. This does not allow for any intra-individual variation in the relationship between exposure and dose.

To determine the total absorbed dose of a chemical agent following a single or multiple occupational or environmental exposures, some knowledge of the toxicokinetics and toxicodynamics of the compound must exist. This information is generally obtained through the use of *in vitro* and *in vivo* animal studies and, in some cases, may be further supported by human volunteer studies. Concentrations of the agent of interest in biological samples such as urine, feces, blood or serum may be used to estimate the total absorbed dose and the accuracy of this dose estimation is based largely on the availability of relevant human studies, and the timing and completeness of sample collection.

The measurement of exposure and dose for compliance, toxicological evaluations and risk assessment generally involves the collection of biological, dermal, or personal or area air samples, and depends on the potential routes of exposure. With the exception of radiation exposure and dose monitoring in occupational settings [9], very rarely is dose estimated in occupational or environmental epidemiologic studies of disease, especially when dealing with chronic occupational exposures. Furthermore, exposure, as previously described, is generally never measured, at least not on an individual basis and not over long periods of observation. The lengthy periods of observation required in epidemiologic studies of chronic disease almost always necessitate the use of surrogate measures of exposure, and in some cases, dose.

Current Methods of Pesticide Exposure Assessment in Epidemiologic Research

The measurement, or more accurately, the evaluation of exposure to pesticides in epidemiologic research is far less sophisticated when compared to risk assessment. In case-control studies of cancer, information on potential exposure to pesticides is most often obtained directly from subjects or their next of kin. In community based studies, the evaluation of pesticide exposure may be as crude as a self-reported or next-of-kin reported yes/no or ever/never classification for broad groups of pesticides. In some circumstances (i.e. rural settings), lifetime histories of specific pesticides used have been obtained and proxy measures of exposure collected, including: earliest and last year of use; number of days a year used; type of spray equipment; use of protective clothing and equipment; frequency of clothing changes; and symptoms of acute pesticide poisoning that may indicate intermittent but high levels of exposure [10,11]. Difficulties in long term recall of pesticide use, differential recall, uncontrolled confounding and the lack of any quantitative exposure or dose measurement, have limited the usefulness of most, if not all, case-control studies in helping to establish cause-effect relationships.

The possibility of accurate pesticide exposure measurement exists in cohort studies. However, in both prospective and retrospective studies, industrial usage records have typically served as surrogate estimates of pesticide exposure and total body dose is assumed to be positively related to the amount used. In the absence of adequate records of pesticide use, job titles and length of employment data have been used, but these methods do not result in quantitative estimates [10]. Past studies on farmers made use of exposure measures such as duration of use, frequency of use, and number of acres sprayed [11-13]. Clearly, these ordinal measures of "use" or surrogates of exposure do not necessarily reflect true continuous "dose" and can result in substantial exposure misclassification.

Furthermore, the use of job titles is likely not representative of actual exposures. In an evaluation of the within- and between-worker variance components of 20,000 occupational exposures, Kromhout *et al.,* found that based on job title and factory, only 25% of the 165 occupational groups in the database had 95% of their individual mean exposures within a two fold range. Approximately 30% of the groups had 95% of individual mean exposures in a range which was greater than 10-fold [14]. In this classic study, it was demonstrated that occupational groups were not uniformly exposed. Thus, to estimate dose or exposure to pesticides, it is important, not only to have access to use records, and job titles, but also to have information about training, spraying practices, and hygiene practices for each individual. In the absence of adequate pesticide use and other relevant records, subjects or their employers may be asked to recall these potentially important determinants of exposure.

Pesticide Use Recall

The validity and reliability of recall of pesticide use is a central issue in epidemiologic research aimed at evaluating the risk associated with pesticide exposure. Johnson *et al.,* compared pesticide data obtained from farmers and their proxy respondents who were interviewed approximately 10 years after an initial interview with the case or control subject [15]. Agreement was quite good for demographic and basic farming information. Generally, the use of proxy information resulted in lower estimates of risk for most pesticides, with a few notable exceptions: any herbicide use, 2,4-D, Alachlor, Atrazine, and DDT. Risk estimates that were based on proxy information were consistently lower for the use of any pesticides, animal insecticides, crop insecticides and fungicides. Assuming that farmers provided accurate pesticide data, this study demonstrated a potential bias with the use of proxy information. However, it did not address the issue of validity of exposure estimation or pesticide use.

In a similar study, Blair and Hoar-Zahm did not observe a reporting bias with the use of proxy information, but found that proxies tended to report pesticide use less frequently [16]. In contrast, another group found potentially serious bias in risk estimates resulting from the use of proxy respondents in the National Cancer Institute's (NCI) Iowa/Minnesota NHL case control study [17]. Using logistic regression and an evaluation of odds ratios for pesticide use, only animal insecticides had consistent risk estimates between proxy and direct informants for three categories of pesticide use frequency: 1-4 days/year, 5-9 days/year and 10 or more days/year. Significant interaction terms between respondent type and pesticide use categories were observed. The proxy-derived odds ratio for 2,4-D use was 2.5 (CI=0.8,8.0) for the highest frequency of use and the direct informant-derived odds ratio was 0.7 (CI=0.3,1.9). The authors conclude that the type of respondent may act as an effect modifier in the association between pesticide exposure and cancer. They legitimately recommend that agricultural pesticide use risk estimates from case control studies should be analyzed and presented separately by the type of respondent [17].

Based on these studies, it appears that in some cases, proxies may be more likely to identify the most commonly used and somewhat controversial pesticides. Furthermore, they may tend to overestimate the frequency of use. This is not surprising when one considers the amount of media attention that has been directed towards some of these compounds. Alachlor, Atrazine, DDT, and 2,4-D have all been associated with adverse environmental or human health effects that have been highly publicised. While there have been a number of studies which have examined the reliability of proxy respondents, no direct studies of the validity of recall of pesticide use in any occupational cohort have been reported [15,16]. It is likely that such studies have not been conducted because of the lack of standardized use records.

To increase the efficiency of case-control studies on agricultural chemical exposures, others have proposed the use of circumstantial determinants of pesticide use [18]. *A priori* exposure matrices based on circumstantial determinants such as crops cultivated, surface area and crop infestations, were incorporated in a case-control study questionnaire and were recalled more frequently than specific chemicals. The proportion of missing values decreased dramatically for the use of specific chemicals, the estimation of dose for each treatment, as well as for the number of treatments a year in a random sample of 40 questionnaires. When the matrix was applied, the number of workers that used specific chemicals such as DDT, parathion and mancozeb increased [18]. It is likely that the use of the matrix could result in increased sensitivity and specificity of pesticide use classification and based on the self reported area and crop infestations, estimates of volumes used may be incorporated into exposure measures. Furthermore, there is little reason to believe that cases would recall these determinants more accurately or frequently than controls, reducing the potential for reporting bias.

Measurement Error in Epidemiologic Studies

Exposure measurement error, recall bias and confounding, may all have important effects on the validity and reliability of estimated risks associated with pesticide or other occupational exposures. In many instances, epidemiologists will assert that exposure measurement error will bias a risk estimate, often the odds ratio (OR), towards the null hypothesis. This axiom is often used to support conclusions of studies in which statistically significant positive risks are observed but at low levels of effect. Authors will argue that the true exposure and disease relationship is likely to be stronger but the observed result is diluted due to exposure measurement error. This traditional treatment of the effects of exposure measurement error is overly simplistic and is based on a number of assumptions, the most conventional being that the exposure measurement error is non-differential. Often, other assumptions necessary to support the conclusion of a potential bias towards the null are overlooked or not appropriately evaluated.

Unfortunately, these widespread and overly simplistic assumptions infiltrate many scientific disciplines that use or interpret the results of epidemiologic studies including those of toxicology and risk assessment. Exposure measurement error in the design, statistical analysis and evaluation of an epidemiologic study needs more sophisticated treatment. Some of the factors to be considered include random and systematic error, differential and non-differential misclassification of the main exposure variable and potential confounders, joint misclassification of exposure and disease, the true level of measurement (continuous versus categorical), and the classification of continuous variables into ordinal or nominal variables [1,19-28]. All these factors may bias risk estimates with seemingly unpredictable results. An understanding of the effects of measurement error is essential in the field of epidemiology and becomes even more crucial in the field of environmental epidemiology, where error in exposure measurement is often the most critical factor limiting the validity and usefulness of study findings.

There is a multitude of possible consequences of exposure measurement error that can result in biased study results. Furthermore, even under the assumption of non-differential error, the power of a study to detect significant health risks will be decreased, in many circumstances, quite significantly [21,29]. The effects of differential or non-differential measurement error in the presence of covariates, or covariates measured with error are even less predictable [1]. When it is considered that most epidemiologic analyses are multivariate, the assumption of bias of risk estimates towards the null due to exposure measurement error seems even less convincing, at least in the absence of any validation data.

Internal or external validation studies have been proposed as a means to evaluate the relationship between the perfectly measured exposures and the imperfect measure of exposure that has been or will be used in an epidemiologic study. In reality, validation studies can only be based on alloyed gold standards when the relevant or true exposure is defined as the biologically effective dose that reaches the target organ (see Figure 1). In practice, the currently accepted technique of dose or exposure measurement will be used to estimate exposure and this will be employed as a gold standard by which to evaluate the imperfect measure. Validation studies may be used to provide estimates of mean exposure to various occupational groups or used to adjust study results. However, the sensitivity of the proposed correction methods to inaccuracies in the assessment of the true exposure may bring their use into question. Since it is generally not possible or cost-effective to obtain accurate measures of dose for all individuals in an epidemiologic study, alternative methods must be developed. Models developed to predict exposure and/or dose for epidemiologic studies provide some promise.

Exposure and Dose Prediction Studies

Very few studies have been designed specifically to develop stochastic or deterministic models to predict current or past exposure or dose in occupational or environmental settings. However, several studies have been conducted to evaluate the determinants of exposure and/or dose, or to evaluate modifiers of the exposure-absorbed dose relationship in both population-based and occupational samples. Although many of these studies are not easily generalizable to other populations, their results help to direct the development of appropriate environmental and biological sampling techniques and the development of exposure-based questionnaires for epidemiologic research. The majority of literature published on the determinants of exposure and dose has focussed on environmental tobacco smoke and lead in both occupational and community based settings. Measures of other occupational exposures have been evaluated and include urinary and red blood cell (RBC) chromium[30], extremely low-frequency magnetic fields (ELF-MF)[31], aromatic hydrocarbons [32], ethylene oxide (ETO) [33] and 2,3,7,8-TCDD [34]. Depending on the type of sample (community-based, child, teenager, adult, occupational, specific industries), internal doses may be quite different and the factors that affect dose will vary by sample. This points to the importance of conducting validation and dose prediction studies within epidemiologic cohorts (internal validation) or within external groups that are known to be closely related with similar potential modifiers of the exposure-dose relationship.

Pesticide Dose and Exposure Prediction Models

Although pesticides are the subject of numerous epidemiologic studies, very little work has been reported on the development of methods or models to improve the prediction of dose over the short or long term. Since many pesticides currently registered for use have short half-lives of elimination measured in hours or days, it is not possible to obtain an estimate of life-time or chronic dose with the use of biological samples. Current dose may be assessed with well-timed biological samples and past exposures may be constructed based on changes in pesticide use over time, application techniques, hygienic conditions, and other factors that may affect absorbed dose. Cumulative dose estimates, which are based on historical dose construction, cannot be validated for these compounds unless specific biomarkers of exposure and/or effect are discovered.

An estimate of lifetime dose based on biological samples is possible for some of the lipid soluble pesticides such as DDT and heptachlor, which have long half-lives (measured in years) either as the parent compound or metabolites. These compounds will be stored in the fat, and in the absence of any significant weight loss, the lipid concentrations can indirectly represent a cumulative body burden when the half-life of excretion is taken into account. Unfortunately, these cumulative dose estimates alone, do not provide sufficient information on the timing, duration, and intensity of exposure that may be important in establishing a cause-effect relationship between pesticide exposures and disease.

Retrospective Dose and Exposure Models

A number of studies have been conducted to evaluate the relationship between chronic exposure to organochlorine compounds and the risk of cancer. Much of this work has focussed on the risk of breast cancer associated with exposure to DDT and polychlorinated biphenyls (PCBs) [35-38], although currently, the focus appears to be changing to case-control studies on the risk of prostate cancer and reproductive effects and cross-sectional studies of endocrine effects. Although a number of studies have been prospective in design, many of these studies have made use of lipid or serum samples collected from subjects participating in case-control studies. The cost associated with these chemical analyses is significant, not to mention the invasiveness of the sampling procedures. Furthermore, the estimates of cumulative body burden obtained from the chemical analyses may not allow for the determination of relevant periods of exposure, their intensity or duration.

In developing countries, where many of these persisting compounds are still used, and where there may be limited access to laboratory analyses, the development of dose prediction models may be necessary for the evaluation of health risks associated with these compounds [39]. The concentration of DDT and its metabolites in abdominal adipose tissue was determined in a random sample of 40 workers selected from 371 malaria control workers in the state of Veracruz, Mexico. Based on occupational history collected by questionnaire, an indirect index of occupational exposure was constructed using job type and duration, and an exposure intensity ranking. Individuals directly involved in the application of the insecticides received the highest ranking. Based on multiple regression analysis, the authors were able to explain 55% of the variation in the concentration of *p,p*-DDE (the major metabolite of DDT) with the use of the exposure index, a variable for the use of protective gear, and a variable to indicate recent weight loss [39].

The retrospective assessment of exposure to phenoxy herbicides, chlorophenols and dioxins has been conducted for manufacturing workers and pesticide sprayers involved in an international cohort study on cancer risk [40]. A deterministic model based on the general source-receptor model, was developed by industrial hygienists for the level of exposure (Li) where:

$$L_i = L_j \, w_e \, w_c \, w_p \, w_o$$

and L_j = the job-related level of exposure, and the mutually independent and multiplicative weighing factors (w) are w_e = emission factor, w_c = contact factor, w_p = personal protection factor, and w_o = other factors. The job-related levels of exposure (L_j) were based on the opinions of industrial hygienists and some limited exposure data. Individuals involved in spraying phenoxy herbicides were ranked the highest (10) together with those involved in the synthesis and finishing of these compounds. Those working in formulation, packing, maintenance, cleaning, and shipping, to name a few, were assigned lower ranks. Cumulative exposure, that did not include the 5 years prior to the inclusion in the nested case-control study, was calculated by summing the level of exposure (L_i) times the duration of the job (D_i) (see Figure 2). Individuals in the same job may have been performing different tasks that result in different levels of exposure, but since no information was available concerning the proportional distribution of tasks within jobs, a common problem, this variation could not be taken into account. Thus, workers within the same jobs were assigned the same job-related level of exposure. Although this represents a loss in precision, the potential validity gained with the use of the deterministic model using ordinal classifications may be substantial when compared to a crude measure of exposure such as duration only. Although the model contains a subjective component, the authors submit that the deterministic model is likely to be more valid and reliable than when exposure estimates are based entirely on subjective assessment by an expert [40].

Discussion

The measurement of pesticide exposures in humans is challenging. In most studies, measures of short term exposures are reported. Most often these samples

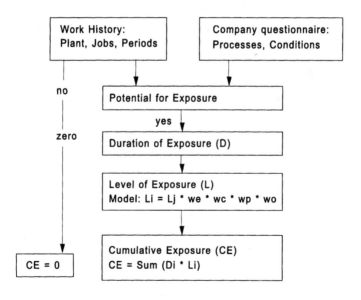

Figure 2. A deterministic pesticide exposure model based on the general source-receptor model. (Adapted with permission from reference 40. Copyright 1994)

are small and not at all representative of the population in which we wish to assess risk. If biological samples are collected, there are issues with the completeness of sample collection (especially if collecting urine), repeated exposures over time, and toxicokinetics and toxicodynamics, making the calculation of absorbed dose difficult. Further, we almost never collect repeated samples within populations over time and have little idea of the variability within individuals as compared with between individuals. Nevertheless, these measurements are used by regulatory agencies to assess acute and chronic risks in humans. Epidemiologic studies, in theory, should help to support these risks assessments. In practice, the data may not be all that useful. Epidemiologic studies designed to evaluate the chronic effects of pesticide exposures are likely severely limited in power, meaning if a true effect exists, we may not be able to detect the effect as being statistically significant. Measurement error of all types can result in reduced study power or biased estimates (up or down) and study results that are difficult to interpret. These results are difficult for epidemiologists to interpret let alone regulators who may not be particularly savvy to the nuances of epidemiologic study design.

Finally, as epidemiologists, we must be able to identify, appreciate and strive to diminish the limitations in our work. We must also attempt to explain these limitations to others who make appropriate and inappropriate use of our results, while stressing the value of such work. Although a recommendation to assess and improve pesticide exposure and dose measurement in epidemiologic research may seem obvious, it is quite remarkable that so little basic research is conducted in this area. In fact, all epidemiologic research could only be improved with this recommendation.

Acknowledgements

This work was funded in part by Health Canada through a National Health Research and Development Program Research Grant (Project No. 6606-5344-502). The author wishes to acknowledge Drs. Andrea Sass-Kortsak, Paul Corey and Jim Purdham for their editorial comments on this review and their time, encouragement, and tremendous support over the years.

References

1. Armstrong, B.; White, E.; Saracci, R. *Principles of exposure measurement in epidemiology*; Oxford University Press: Oxford, 1992; Vol. 21.
2. Munro, I.; Carlo, G.; Orr, J.; Sundi, K.; Wilson, R.; Kennepohl, E.; Lynch, B.; Jablinske, M.; Lee, N. *J Am Col Toxicol.* 1992, *11*, 559-664.

3. Bond, G.; Rossbacher, R. *Br J Ind Med* **1993**, *50*, 340-348.
4. Monson, R. *Occupational Epidemiology*; 2 ed.; CRC Press Inc.: Florida, 1990.
5. Hatch, M.; Thomas, D. *Environ Health Perspect* **1993**, *101 Suppl 4*, 49-57.
6. Moody, R.; Wester, R.; Melendres, J.; Maibach, H. *J Toxicol Environ Health* **1992**, *36*, 241-250.
7. Maibach, H.; Feldman, R.; Milby, T.; Serat, W. *Arch Environ Health* **1971**, *23*, 208-211.
8. Harris, S.; Solomon, K. *J Toxicol Environ Health* **1992**, *36*, 233-240.
9. Muirhead, C. R.; Boice JD, J. r.; Raddatz, C. T.; Yoder, R. C. *Health Physics* **1996**, *70*, 645-650.
10. Blair, A.; Hoar Zahm, S.; Cantor, K.; Stewart, P. In; American Chemical Society: Washington, 1989; Vol. 382, pp 38-46.
11. Zahm, S.; Weisenburger, D.; Babbitt, P.; Saal, R.; Vaught, J.; Cantor, K.; Blair, A. *Epidemiology* **1990**, *1*, 349-356.
12. Hoar, S.; Blair, A.; Holmes, F.; Boysen, C.; Robel, R.; Hoover, R.; Fraumeni JF, J. *JAMA* **1986**, *256*, 1141-1147.
13. Wigle, D.; Semenciw, R.; Wilkins, K.; Riedel, D.; Ritter, L.; Morrison, H.; Mao, Y. *J Natl Cancer Inst* **1990**, *82*, 575-582.
14. Kromhout, H. E.; Symanski, E.; Rappaport, S. *Annal Occup Hygiene* **1993**, *37*, 253-270.
15. Johnson, R.; Mandel, J.; Gibson, R.; Mandel, J.; Bender, A.; Gunderson, P.; Renier, C. *Epidemiology* **1993**, *4*, 157-164.
16. Blair, A.; Zahm, S. *Epidemiology* **1993**, *4*, 55-62.
17. Olsen, G.; Bodner, K. *J Agromed* **1996**, *3*.
18. Nanni, O.; Ricci, M.; Lugaresi, C.; Amadori, D.; Falcini, F.; Buiatti, E. *Scand J Work Environ Health* **1993**, *19*, 191-199.
19. Flegal, K.; Brownie, C.; Haas, J. *Am J Epidemiol* **1986**, *123*, 736-751.
20. Flanders, W.; Drews, C.; AS, K. *Epidemiology* **1995**, *6*, 152-156.
21. Armstrong, B. *Am J Epidemiol* **1990**, *132*, 1176-1184.
22. Birkett, N. *Am J Epidemiol* **1992**, *136*, 356-362.
23. Correa-Villasenor, A.; Stewart, W.; Franco- Marina, F.; Seacat, H. *Epidemiology* **1995**, *6*, 276-281.
24. Dosemeci, M.; Wacholder, S.; Lubin, J. *Am J Epidemiol* **1990**, *132*, 746-748.
25. Rosner, B. *Stat Med* **1996**, *15*, 293-303.
26. Wacholder, S. *Epidemiology* **1995**, *6*, 157-161.
27. Flegal, K.; Keyl, P.; Nieto, F. *Am J Epidemiol* **1991**, *134*, 1233-1244.
28. Delpizzo, V.; Borghesi, J. *Int J Epidemiol* **1995**, *24*, 851-862.
29. Armstrong, B. *Am J Epidemiol* **1996**, *144*, 192-197.
30. Bukowski, J.; Goldstein, M.; Korn, L.; Johnson, B. *Arch Environ Health* **1991**, *46*, 230-236.

31. Abdollahzadeh, S.; Hammond, S.; Schenker, M. *Am J Ind Med* **1995**, *28*, 723-734.
32. Lagorio, S.; Forastiere, F.; Iavarone, I.; Vanacore, N.; Fuselli, S.; Carere, A. *Int J Epidemiol* **1993**, *22 Suppl 2*, S51-56.
33. Hornung, R.; Greife, A.; Stayner, L.; Steenland, N.; Herrick, R.; Elliott, L.; Ringenburg, V.; Morawetz, J. *Am J Ind Med* **1994**, *25*, 825-836.
34. Ott, G.; Messerer, P.; Zober, A. *Int Arch Occup Health* **1993**, *65*, 1-8.
35. Krieger, N.; Wolff, M.; Hiatt, R.; Rivera, M.; Vogelman, J.; Orentreich, N. *J Natl Cancer Inst* **1994**, *86*, 589-599.
36. Lopez-Carrillo, L.; Blair, A.; Lopez-Cervantes, M.; Cebrian, M.; Rueda, C.; Reyes, R.; Mohar, A.; Bravo, J. *Cancer Res* **1997**, *57*, 3728-3732.
37. Unger, M.; Kiaer, H.; Blichert-Toft, M.; Olsen, J.; Clausen, J. *Environ Res* **1984**, *34*, 24-28.
38. vanqt Veer, P.; Lobbezoo, I.; Martin-Moreno, J.; Guallar, E.; Gomez-Aracena, J.; Kardinaal, A.; Kohlmeier, L.; Martin, B.; Strain, J.; Thamm, M.; van Zoonen, P.; Baumann, B.; Huttunen, J.; Kok, F. *BMJ* **1997**, *315*, 81-85.
39. Rivero-Rodriguez, L.; Borja-Aburto, V.; Santos-Burgoa, C.; Waliszewskiy, S.; Rios, C. *Environ Health Perspectives* **1997**, *105*, 98-101.
40. Kauppinen, T.; Pannett, B.; Marlow, D.; Kogevinas, M. *Scand J Work Environ Health* **1994**, *20*, 262-271.

Chapter 13

Dose Prediction Modeling for Epidemiologic Assessment of Pesticide Exposure Risks in Occupational Cohorts

Shelley A. Harris[1,2,*] and Kristen M. Wells[1,2]

[1]Center for Environmental Studies, Virginia Commonwealth University
[2]Department of Epidemiology and Community Health, P.O. Box 843050,
Virginia Commonwealth University, Richmond, VA 23284–3050
[*]Corresponding author: telephone: (804) 828–1582, fax: (804) 828-1622, and
email: saharris@vcu.edu

Epidemiologic studies designed to evaluate the effects of commonly used turf pesticides may have limited power to detect health risks and may be subject to bias from exposure measurement error. To increase the accuracy and precision of dose estimation for both risk assessment and epidemiologic research, valid models must be developed. Further, repeated measures of exposures over time are necessary to estimate both inter- and intra-individual variation. To address some of these issues, a national study of TruGreen Chemlawn employees was initiated in 2003. In the pilot study, conducted in Richmond, Virginia, up to 19 days of 24-hour urine samples were collected from 22 individuals. In 2004, urine samples were collected from a total of 113 volunteers in the spring, summer and fall, from 5 locations across the United States. The design of this study, the selection of national locations and pesticides, urine sampling methodology, and statistical modeling efforts are described.

Most epidemiologic studies are limited by the lack of valid pesticide exposure data, or more correctly, absorbed dose data. One way to quantify dose is to use biological monitoring techniques that measure urinary concentrations of pesticides. However, for prospective studies with large cohorts following individuals over many years, this is highly impractical and, for retrospective studies, it is not possible. Although the accurate measurement of exposure or dose in prospective cohort studies is theoretically possible, it is practically very difficult. The cost, time commitment, and feasibility of enrolling subjects to provide long-term biological samples are insurmountable. Thus, prospective cohort studies, conducted to evaluate chronic effects of occupational exposures, often rely on many of the same exposure estimation techniques employed in retrospective studies. These techniques may not provide information of sufficient quality to improve our state of knowledge. New methods of dose estimation must be developed specifically for cohorts that are occupationally exposed. Workers employed as professional applicators provide a unique opportunity to develop these methods.

Epidemiologic Studies of Pesticide Applicators

Researchers at the National Cancer Institute in Maryland are following a cohort of approximately 40,000 Chemlawn workers (now called TruGreen Chemlawn) employed as professional turf applicators in the United States. This prospective mortality study, which has a retrospective component, will make use of semi-quantitative estimates of pesticide use such as the number of days worked per year and will be based on the branch where the employee worked, the amount of pesticide purchased for the branch, the period of employment, his/her job title, and the pesticide application program offered at the branch. Individual pesticide use or exposure cannot be estimated due to the lack of records for each individual employee [1,2]. The NCI has recently reported on the retrospective component of the cohort study and, although the cohort was young with a short duration of employment and a short period of follow-up, a significant increase in non-Hodgkin's Lymphoma (NHL) was observed among professional turf applicators employed for three or more years (Standardized Mortality Ratio (SMR) = 7.11, CI = 1.8, 28.4) [3]. As expected, due to the healthy worker effect, which can be described as the phenomenon where cohorts of employed individuals exhibit lower death rates than the general population due to the fact that severely ill individuals are often excluded from employment, the cohort had significantly decreased mortality as compared to the US population from the combined all causes of death. Overall, there were 45 cancer deaths (59.6 expected, SMR = 0.76). Mortality from bladder cancer was significantly increased, but only one subject reported direct occupational contact with pesticides. Overall, there were four deaths due to NHL (SMR = 1.14) and three were male lawn applicators (SMR = 1.63). The authors conclude that the NHL excess is consistent with several earlier studies, but may be due to chance. If adequate measures of exposure or preferably dose can be developed for this cohort, the continued follow-up of these employees presents an excellent opportunity to evaluate health risks associated with some of the commonly used turf pesticides.

Pesticides Commonly Used in the Turf Industry

The phenoxy herbicide 2,4-dichlorophenoxyacetic acid (2,4-D) and other related herbicides have been used extensively in the professional turf industry and are the subject of considerable investigation. Many epidemiologic studies have been conducted and are currently underway to evaluate the chronic effects of these pesticides in occupational groups, and although the current weight of epidemiological evidence may be suggestive of an association between the use of 2,4-D and some cancers, a cause-effect relationship has not been established. A number of reviews concerning exposure and the possible health effects of 2,4-D, dicamba (benzoic acid herbicide, 3,6-dichloro-2-methoxybenzoic acid), and related phenoxy herbicides such as mecoprop (2-(2-methyl-4-chlorophenoxy) propionic acid, MCPP; phenoxypropionic herbicides) and their dioxin contaminants are available in the literature. Until quite recently (2000), chlorpyrifos was used extensively for insect control in the turf industry. By many companies, it has now been replaced by the pyrethroid insecticide bifenthrin (2-methyl-1,1-biphenyl-3-y1)-methyl-3-(2-chloro-3,3,3-trifluoro-1-propenyl)-2,2-dimethyl cyclopropanecarboxylate) and the chloro-nicotinyl insecticide imidacloprid (1-[(6-chloro-3-pyridinyl)methyl]-N-nitro-2-imida-zolidinimine)).

No information is available on the exposure or dose of professional turf applicators to the chlorpyrifos replacement insecticides imidacloprid and bifenthrin. Considering the extensive use throughout North America, it is surprising that only one study of applicator exposure to imidacloprid (while spraying mangoes) has been published in the peer-reviewed literature [4]. Similarly, we found only one study published on bifenthrin exposure [5] and one with pest control operators (PCOs) to the pyrethroid cyfluthrin [6,7]. No studies on fungicide biomonitoring in professional turf applicators were found in the literature search.

Pesticide Exposure and Dose Prediction Modeling

The methods used to predict pesticide exposure or dose following occupational or environmental exposures depend largely on the intended use of the information. Models that have been developed to evaluate pesticide exposures for registration or re-registration purposes are generally based on worst-case scenarios and are designed to apply across all individuals.

Assumptions may include 100% absorption of a pesticide through skin and constant body weight and breathing rates for all individuals. These deterministic models do not typically allow for individual variation; are generally based on the estimation of exposure, not dose; are usually conducted under experimental, not observational settings; and are most often designed to present worst-case estimates of exposure.

Under experimental settings, deterministic pesticide dose prediction models have been developed for the purpose of assessing uptake of pesticides from contaminated turf and these models have been evaluated [8,9]. Estimates of potential bystander exposure to several herbicides have been made to establish re-entry intervals for product registration in Canada [10]. Although necessary for risk assessment and product registration, these types of models may only be useful for the semi-quantitative estimation of environmental exposure to pesticides in domestic settings. They do not allow for the individual prediction of dose, and do not account for multiple sources of exposure.

A number of predictive models have also been developed to estimate agricultural exposure to pesticides for registration purposes [11]. For example, the Pesticide Handlers Exposure Database provides exposure information for mixers, loaders, applicators, and flaggers, under a number of environmental, hygienic (protective clothing), and working conditions (type of spray equipment) [12]. This information can aid in the construction of prospective and historical exposures for individuals in epidemiologic studies, if appropriate information on glove use, protective clothing worn, and application procedures is collected over time. Biological validation studies using this database are important so that factors affecting total body dose can be evaluated [11]. This would allow for an evaluation of the assumptions used in exposure assessment and could provide estimates of the individual variation in dose relative to potential exposure. In the absence of this type of information, similar exposure groups can be defined but the relationship between potential exposure and dose is unknown, as is the variation in dose within these groups.

When the goal of pesticide dose prediction is for the improvement or evaluation of exposure assessment methods for epidemiologic research, the approach is somewhat different to that used for registration or risk assessment purposes. Instead of attempting to predict dose with the use of exposure information, individual dose is measured (or estimated) with the use of biological samples and information is collected to evaluate the factors that influence dose. One excellent example of this type of work is the validation studies underway for the Agricultural Health Study (AHS). The National Cancer Institute (NCI), the National Institute of Environmental Health Sciences (NIEHS), and the U.S. Environmental Protection Agency have initiated the AHS, which is being conducted in Iowa and North Carolina. As part of this large cohort study (75,000 adults), detailed monitoring of pesticide dose will be conducted on 200 families, and investigators will attempt to relate the internal dose to pesticide application procedures and protective practises, and account for both direct and indirect exposures. The questionnaire information will also be supplemented with data from the Pesticide Handlers Exposure Database [13]. Other researchers at the University of Minnesota have conducted a comprehensive Farm Family Exposure Study to evaluate factors associated with exposure to 2,4-D, glyphosate, and chlorpyrifos and absorbed dose [14]. Reporting of results is underway.

In Canada, researchers at the University of Guelph and from Health Canada are conducting a farmer dose evaluation study of approximately 300 farm families. In this study, the internal dose of 2,4-D will be evaluated for the farm operator, the spouse, and one child in the family. Factors contributing to internal dose such as contamination of drinking water, drift of chemicals, and the use of personal protection devices will be evaluated [15].

A number of studies have been conducted to evaluate the exposure of professional turf applicators to pesticides, including the herbicides 2,4-D, MCPA (4-chloro-2-methylphenoxyacetic acid), Mecoprop (2-(4-chloro-2-methyl phenoxy) propionic acid) (MCPP), and Dicamba (3,6-dichloro-O-anisic acid), a benzoic acid compound [16-18]. In a large sample of 98 professional turf applicators from 20 companies in Southern Ontario, daily dose estimates of 2,4-D ranged from 0.004 to 19 mg/day with a geometric mean of 0.42 mg/day [16]. Doses of mecoprop were consistently higher and ranged from 0.006 to 23 mg/day with a geometric mean of 0.584 mg/day. Individuals who sprayed pesticides only had the highest average doses in the study and, contrary to current thinking, those who were involved in spraying and mixing had, on average, lower doses. Those who only mixed pesticides during the week of the exposure study had the lowest doses in the study. Based on job titles, applicators had the highest absorbed dose, followed by owners of the companies and managers. Again, current thinking would have predicted that owners receive the lowest doses. Since these workers were repeatedly exposed to varying amounts of pesticides, a method of dose estimation was developed to predict total weekly dose that would allow for different use patterns by each individual [19]. Further, since accuracy of dose estimates is dependent on the collection of 24-hour urine samples, both creatinine excretion and self-reported missed samples were used to evaluate collection completeness [20]. During a one week period, the volume of pesticide applied was weakly related to the total dose of 2,4-D absorbed ($R^2=0.21$) [16]. Two additional factors explained a large proportion of the variation in dose: the type of spray nozzle and the use of gloves. Job satisfaction and current smoking influenced the dose but were not highly predictive. In the final multiple regression models predicting total absorbed dose of 2,4-D and mecoprop, 63 to 68 percent of the variation was explained. Commonly used job titles and duties performed explained only 11 and 16 percent of the variation in dose, respectively [21]. The amount of pesticide sprayed over the work week was more predictive of dose than the use of job titles or tasks performed [21].

The future application of these Canadian models for epidemiologic research will depend on their external validity, availability of information and records from employers, the feasibility of contacting study subjects, and cost. Clearly, an external validation of this model for use in epidemiologic studies of professional turf applicators in the United States is desirable, given the lack of consensus concerning the carcinogenic and/or reproductive effects of many commonly used pesticides in the current epidemiologic literature, and the lack of studies on the chronic hazards associated with exposure to some of the newer replacement compounds.

Measurement Issues in Epidemiologic Studies

An overview of current methods of pesticide exposure assessment and measurement error issues is provided in a companion chapter (Harris, 2005) in this Series. In occupational epidemiologic studies of pesticide exposures it has generally been assumed that industrial usage records serve as a surrogate estimate of pesticide exposure and total body dose is assumed to increase as the amount used increases. However, in the absence of adequate records of pesticide use, job titles and length of employment data have been used as proxies of exposure, but these do not result in quantitative estimates of dose [2].

Previous work demonstrates that in professional turf applicators an estimate of dose based on pesticide use records will result in a substantial exposure misclassification [16,17,21]. Therefore, estimating use based on pesticide purchase (as proposed in the NCI cohort study), may result in even greater misclassification of exposure. If absorbed dose estimates are based on pesticide use data (a proxy for exposure) for epidemiologic studies, the sample size necessary to detect a significant health effect would be close to six times higher than if the perfectly measured dose was used [16]. If the recent research on the relationship between pesticide use and resulting dose is generalizable to professional turf applicators as a whole, this presents serious implications for the effectiveness of current studies to detect any statistically significant association between pesticide exposure and adverse health effects.

Objectives

The primary objective of this study was to obtain repeated measurement of pesticide exposures within individuals over time, to evaluate the factors associated with absorbed dose, and to validate previously developed dose prediction models in a national sample of TruGreen Chemlawn workers. The design, execution, and proposed statistical models are described.

Methods and Materials

The graphical representation of the study design is presented in Figure 1. The pilot study, conducted in 2002, was designed to obtain repeated 12- or 24-hour urine samples from 10 individuals over a 5-day period in the summer and over a two week period in the fall. These periods were timed to coincide with the heaviest spraying of insecticides (bifenthrin, imidacloprid) and herbicides (MCPA, mecoprop, and dicamba) at the Richmond, Virginia branch. The comprehensive evaluation of the urinary excretion of these pesticides was designed to obtain information on the toxicodynamics and toxicokinetics of these pesticides in humans following repeated exposures.

Following a verbal presentation to the Richmond branch, recruitment of volunteers was much greater than expected and requests for the tree and shrub applicators to participate were received. Thus, the study was expanded to include these employees in addition to the turf applicators. A total of 22 workers signed informed consent to participate and complete samples (19 days of 24 hours urine samples) were obtained from 12 individuals. Subjects were paid $10 a sample, for a total of $190 if they completed the entire study.

Development of Worker Exposure Questionnaire

A previously developed questionnaire was revised to include information relevant to insecticide use and exposure for pilot testing in 2003. It was designed to measure all known variables that could potentially increase or decrease pesticide exposure in relation to the amount handled, with a focus on dermal absorption. Potential factors include: age/sex; smoker/non smoker; length of training; licensed/non-licensed; number of years employed/licensed; pesticide formulation (granular vs. liquid); type of spray equipment used (i.e. injection, high or low pressure nozzles); mixing/filling duties; protective equipment worn (gloves, overalls, rubber boots, etc.); occurrence of spills during mixing, application, etc.; frequency of uniform laundering; and personal hygiene (washing prior to lunch, etc.). Based on some of the previous work by Slovic and others [22-24], questions on risk and risk/benefit perceptions were developed and questions to elicit self-reported exposures were formulated. The revised questionnaire was tested in the group of 22 workers from the Richmond, VA branch in the summer and this resulted in minor changes in question numbering and skip patterns. The questionnaire was revised and given each Friday (i.e. 2 more times for each volunteer) during the fall herbicide monitoring study.

Year 1

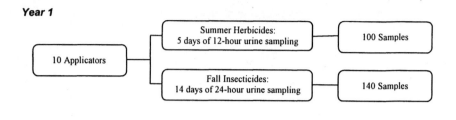

Year 2

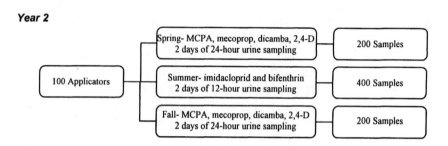

Figure 1. TruGreen Chemlawn Dose Monitoring Study Design

National Study

Branch Selection

Due to the seasonal and often short-term nature of employment in the lawn care business, initial contact with subjects in 2004 was through five TruGreen Chemlawn branches. For the National Study, we planned to sample five branches and/or franchises to reflect national differences in pesticide programs and timing of applications. The 5 geographic locations chosen for National Study were: 1) Northeast; 2) Southeast; 3) Northwest; 4) Southwest; and 5) Central.

To participate in the National Study, all locations were required to use both insecticides and herbicides and needed to be of sufficient size to obtain a total sample size of 100 employees (i.e. approximately 20 employees per branch). Logistically, we needed to select sites so that our sampling season was as long as possible (April to November, 2004) and that the fieldwork could be conducted with a small number of staff. Further, direct or 1-stop flights from Richmond/ Dulles/ Williamsburg area were preferred. Pesticide use data (all pesticide products and monthly use for all locations in the US and Canada) was provided by the TruGreen Chemlawn corporate office. After careful consideration of each branch's pesticide use patterns, particularly the use of both insecticides and herbicides throughout the year, location, and length and intensity of each insecticide and herbicide spray season, a list of 17 branches representing six of the company's eight regions was presented by study personnel to TruGreen Chemlawn corporate management (see Figure 2). After consideration of the number of employees at each branch and the potential level of cooperation of each branch, 5 TruGreen Chemlawn branches were selected by TruGreen Chemlawn corporate managers for inclusion in the study: Sterling, Virginia (D.C. West); Plano, Texas; Puyallup, Washington; Plainfield, Illinois; and Salt Lake City, Utah.

Subjects and recruitment

Following final approval of study locations, individual branch and operations managers (generally two to four at each location) were contacted and group meetings with applicators were arranged for the spring sampling period (see Figure 1). Potential participants were given both oral and written information on study background, aims, and procedures, and the 113 employees from the five locations willing to participate provided signed consent. General inclusion criteria included being at least 18 years of age and having potential contact with pesticides as part of the employee's job description. At two study locations, however, the inclusion criteria were modified at the request of the branch operations managers. In Puyallup, Washington, employees in training were not entered into the study, and Plainfield, Illinois, only herbicide applicators were enrolled. Subjects included both licensed and non-licensed pesticide applicators and were remunerated with $20 per sampling week ($60 total for completion of all 3 seasons) for their contribution to the study. In addition, each subject was allowed to keep the soft-sided cooler bag and ice packs used during sample collection to keep urine samples cold.

A summary of the volunteer enrolment and retention is presented in Table I. Volunteers were actively recruited in the spring and summer (new hires). We were able to visit all national sites three times, except for Plainfield, IL, where logistics prevented a third visit. Retention of study subjects was excellent and dropout was primarily due to layoff or termination of employment.

Sample Collection and Analysis

For the nationwide study, urine samples were collected during three different spraying seasons: the spring (April and May) and fall (October and November) herbicide sprays and summer insecticide spray (June and July). Each study participant was provided with one 3 L urine collection container (Simport

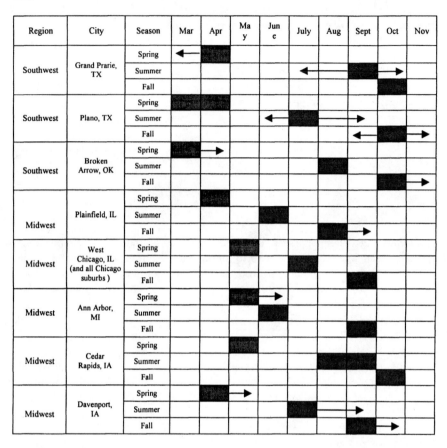

Figure 2. Optimal months (2003) for biological sample collection in select US cites in the Southwest and Midwest. Arrows indicate semi-optimal months.

Plastics Limited, Fisher catalogue number 14-J75-116) for each collection interval, a soft-sided cooler bag, and 2 frozen ice packs. Total urine output was collected for two consecutive 24-hour periods (herbicide) or four consecutive 12- hour periods (insecticide) following a minimum of 3 consecutive workdays. Subjects were asked to store all samples in their cooler bag with ice packs or in the refrigerator when possible during each collection period, and study personnel retrieved and processed samples at the end of each 24-hour interval. Upon collection, samples were visually observed for any inconsistencies in appearance or volume, total sample volume was recorded, and specific gravity was measured using the Leica AR200 digital hand-held refractrometer (Leica catalogue number 13950000). Each sample was divided into three 40 mL aliquots (in 50 mL Corning graduated plastic tubes, Corning catalogue number 430828) and two 100 mL aliquots (in 125 mL Nalgene rectangular HDPE bottles, Nalgene catalogue number 2007-0004), packaged in accordance with Federal dangerous goods shipment guidelines, and overnight shipped in insulated diagnostic shippers (Saf-T-Pak item STP-320) with frozen ice packs and ice blankets to Virginia Commonwealth University. Upon arrival, samples were immediately frozen and were stored at -20° C until analysis.

To evaluate completeness of urine collection, one 40 mL aliquot from all 24-hour urine samples was analyzed for creatinine content by Scientific Testing Laboratories (Richmond, Virginia). If necessary, urine volumes will be corrected for self-reported missed sample collection [20]. Urine samples are currently undergoing analysis for MCPA, mecoprop, bifenthrin metabolites, imidacloprid metabolites, dicamba, and 2,4-D using solid-phase extraction followed by positive/negative ion electrospray ionization HPLC/MS/MS; a method developed as part of the project in the Chemical Response and Terrorism Preparedness Laboratory at the Virginia Department of Consolidated Laboratory Services (Richmond, Virginia). This method is capable of quantifying levels of all analytes to 1 part per billion (1.0 µg/L) [25]. Although the study was originally designed to evaluate herbicides and insecticides separately (i.e. a focus on herbicides in the spring and fall and insecticides in the summer, see Figure 1) the development of a method to simultaneously measure all analytes will provide much more useful data on individual variability.

Information Obtained from Employers

Daily pesticide use records (volume used and area sprayed) for each subject in the pilot and national studies have been obtained from the operations

Table I: Summary of Nationwide Study Subject Enrolment

City	Subjects Enrolled	Spring Completed/ Enrolled	Summer Completed/ Enrolled	Fall Completed/ Enrolled
Sterling, VA	33	28/31	19/31	22/33
Plano, TX	14	14/14	14/14	14/14
Puyallup, WA	19	13/13	17/17	11/19
Salt Lake City, UT	27	22/22	19/27	15/27
Plainfield, IL	20	20/20	15/20	N/A
Total	113	97/100	84/109	62/93

managers at each location. These records will be used to compare with self-reported employee exposures and as a gold standard for estimating dose with the urinary concentration data.

Discussion

Given the lack of consensus concerning the carcinogenic effects of 2,4-D and other pesticides, the lack of absorbed dose information for some of the pyrethroids, and the more recently introduced insecticide (imidacloprid) in the current epidemiologic and toxicologic literature, professional turf applicators are an important cohort of workers for future epidemiologic studies. Their exposures will likely exhibit the substantial variation necessary to establish dose-response relationships. Moreover, their exposures to other biological factors, chemicals, and other pesticides, which could confound relationships, will generally be less than that of other occupational groups, such as farmers.

The results of this study will help to refine pesticide dose prediction for both epidemiology and risk assessment. The design of the study, to include repeated measurements in individuals over time, will allow for the evaluation of variability in exposures over time, for individuals and different pesticides. This work is unique. Moreover, pesticide exposure prediction models can be developed and tailored specifically as the available information permits. For example, if it is impossible to contact individual employees in a retrospective study, an exposure prediction model that includes information available from employers such as number of training days, size of business, number and gender of employees, along with individual or group pesticide use records would be more predictive of individual/group exposures than a model containing measurement of pesticide use alone. If however, it is possible to contact individual employees, the accuracy of prediction of individual exposure is likely to be increased through the ability to add a few more important predictor variables.

Future dose prediction model efforts ultimately depend on the proposed use of the model but currently include the development of a statistical model that:

1) Predicts total weekly dose and may include any of the explanatory variables that have been evaluated in this study. This model will be most useful for long-term monitoring in the industry and determining effective abatement strategies.
2) Will be restricted to allow only the inclusion of predictor variables on which information can be collected from contact with the employers and access to their records. This model will be most useful in longer-term or short-term retrospective studies in which it is not possible to contact individual subjects or the validity or reliability of the information collected from the subjects is questionable.

3) Will be restricted to allow only the inclusion of explanatory variables on which information can be collected from contact with the employers, access to their records, and contact with the individual subjects, while considering the logistics of collecting the information

A comparison of models 2 and 3 with model 1 will allow for the evaluation of the potential misclassification of exposure (differential and non differential) when it is not possible to collect all the information that will be obtained in the current study as reported in this chapter. In addition, models 2 and 3 may be directly applied to estimate dose in a prospective or retrospective epidemiologic study with the added benefit of some understanding of exposure misclassification. This allows for an upward adjustment of sample size in the design phase of any future proposed studies or a calculation of the potential bias of risk estimates in current studies. Although we will base our models (1, 2 or 3) on estimation of weekly dose of individual applicators, we believe that the model(s) can be adjusted to estimate seasonal and lifetime dose, which will be more useful for the evaluation of chronic health effects associated with pesticides exposures.

Acknowledgements

This work was funded in part by a grant from CDC/NIOSH, PA-99-143, Occupational Safety and Health Research, R01 OH004084. The authors wish to acknowledge the contribution of numerous individuals involved in the project described, including: Kirk Hurto, Chis Forth, Roger Yeary, Ross Eckstein, Scott Fairchild, David Patterson, Scott Ericksen, Emil Heihn, Scott Roberts, Rick Pearson, Dave Crago, Ray Rudie, Jim Hanson, Brian Hoxie, John Tiefel, Byron Smith, Tony Wilmes, Brad Brown, Tom Curry, Mike Palermo, and Todd Evan (TruGreen Chemlawn); Charlene Crawley, J. Clifford Fox, James Mays, and Tim Croley (VCU); Ricky Ciner, Jean Nelson, and Donna Huggins (graduate students, VCU); and, of course, the motivated and cooperative volunteers.

References

1. Hoar Zahm, S. "Mortality study of Chemlawn employees: A retrospective and prospective study. Protocol.," National Cancer Institute, Occupational Studies Section, 1987.
2. Blair, A.; Hoar Zahm, S.; Cantor, K.; Stewart, P. In; American Chemical Society: Washington, 1989; Vol. 382, pp 38-46.
3. Zahm, S. *J Occup Environ Med* 1997, *39*, 1055-1067.
4. Calumpang, S. M.; Medina, M. J. *Bull Environ Contam Toxicol* 1996, *57*, 697-704.

200

5. Smith, P. A.; Thompson, M. J.; Edwards, J. W. *J Chromatogr B Analyt Technol Biomed Life Sci* **2002**, *778*, 113-120.
6. Leng, G.; Kuhn, K. H.; Idel, H. *Toxicol Lett* **1996**, *88*, 215-220.
7. Wieseler, B.; Kuhn, K.; Leng, G.; Idel, H. *Bull Environ Contam Toxicol* **1998**, *60*, 837-844.
8. Durkin, P.; Rubin, L.; Withey, J.; Meylan, W. *Toxicol Ind Health* **1995**, *11*, 63-79.
9. Harris, S.; Solomon, K. *J Environ Sci Health [B]* **1992**, *27*, 9-22.
10. Harris, S. A.; Solomon, K.; Stephenson, G.; Bowhey, C. "The Use of Dislodgeable Residue Data of Triclopyr and Clopyralid from Turf to Estimate Potential Human Exposure.," Canadian Centre For Toxicology. Report: Green Cross, A Division of FISONS. 6050 Century Ave., Mississauga, Ont., 1991.
11. van Hemmen, J. *Rev Environ Contam Toxicol* **1992**, *128*, 43-54.
12. PEHD. Pesticide Exposure Handlers Database, Reference Manual. Springfield:Versar Inc., 1992.
13. Alavanja, M.; Sandler, D.; McMaster, S.; Zahm, S.; McDonnell, C.; Lynch, C.; Pennybacker, M.; Rothman, N.; Dosemeci, M.; Bond, A; Blair, A. *Environ Health Perspect* **1996**, *104*, 362-369.
14. Acquavella J. F.; Alexander, B. H.; Mandel J. S.; Gustin, C.; Baker, B.; Chapman, P. *Environ Health Perspect* **2004**, 112, 321-326.
15. Ritter, L. In *CNTC Announcements*; Canadian Network of Toxicology Centers: Guelph, 1996.
16. Harris, S. A.; Sass-Kortsak, A. M.; Corey, P. N.; Purdham, J. T. *J Exposure Anal Environ Epidemiol* **2002**, *12*, 130-144.
17. Solomon, K.; Harris, S.; Stephenson, G. In *Pesticides in Urban Environments.*; 522 ed.; Racke, R., Leslie, A., Eds.; American Chemical Society: Washington, 1993; pp 262-274.
18. Yeary, R. *Appl Ind Hygiene* **1986**, *3*, 119-121.
19. Harris, S. A.; Corey, P. N.; Sass-Kortsak, A. M.; Purdham, J. T. *Int Arch Occup Environ Health* **2001**, *74*, 345-358.
20. Harris, S. A.; Purdham, J. T.; Corey, P. N.; Sass-Kortsak, A. M. *AIHAJ* **2000**, *61*, 649-657.
21. Harris, S. A.; Sass-Kortsak, A. M.; Corey, P. N.; Purdham, J. T. *Am J Ind Med* **2005**.
22. Slovic, P. *Risk Anal* **1999**, *19*, 689-701.
23. Slovic, P.; Malmfors, T.; Mertz, C.; Neil, N.; Purchase, I. *Hum Exp Toxicol* **1997**, *16*, 289-304.
24. Slovic, P. *Science* **1987**, *236*, 280-285.
25. Ciner, F.; Croley, T. R.; Harris, S. A.; Crawley, C. D. *American Chemical Society, 229th National Meeting, Abstract* **2005**, Spring 2005,79.

Chapter 14

Evaluation of Potential Carbaryl Exposures Associated with Residential Lawn and Garden Product Use: Results of a Biological Monitoring Program

Curt Lunchick[1], Jeffrey H. Driver[2], and John H. Ross[3]

[1]Bayer CropScience, 2 T.W. Alexander Drive, Research Triangle Park, NC 27709
[2]infoscientific.com, Inc., 10009 Wisakon Trail, Manassas, VA 20111
[3]infoscientific.com, Inc., 5233 Marimoore Way, Carmichael, CA 95608

Bayer CropScience conducted a biological monitoring study of individuals residing in homes in California or Missouri in which carbaryl was applied by a member of the family. The study was designed to monitor the absorbed dose of carbaryl in an adult applicator residing in each of the homes and all occupants of the home aged four years or older. The study did not control the post-application activities of the study participants and provides a comparison to residential exposure estimates based on standardized routines such as Jazzercise® (1, 2). Key findings from the study were as follows: 1) The exposure to applicators applying carbaryl by hose-end spray applicator were comparable to independent measurements using similar spray equipment conducted by the Outdoor Residential Exposure Task Force; 2) Pre-application 1-naphthol urine levels were comparable to 1-naphthol levels reported from the US EPA's National Human Exposure Assessment Survey (http://www.epa.gov/nerL/research/nhexas/nhexas.htm); 3)The 4-12 year age group had the highest overall exposure of all cohort groups in the study; 4) Yard activity was the primary determinant for post-application exposure potential; 5) Application rate played a less significant role in post-application exposure than the post-application contact with treated areas; 6) Secondary routes of exposure such as vapor intrusion, track-in, or dust levels appear to be insignificant sources of exposure; and 7) The US EPA's Office of Pesticide Program's screening-level assessment of potential residential exposure to carbaryl overestimates actual monitored exposure levels because of overestimation biases associated with the assumed amount of active ingredient handled and the assumed lawn re-entry activity patterns.

Introduction

Carbaryl is a N-methyl carbamate insecticide that is currently undergoing re-evaluation by the U.S. Environmental Protection Agency (EPA). As part of the ongoing reregistration process in the United States (US), Bayer CropScience, formerly Aventis CropScience, conducted a biological monitoring study of individuals residing in homes in California or Missouri in which carbaryl was applied by a member of the family. The carbaryl product was Sevin® GardenTech Ready-To-Spray containing 22.5% carbaryl as the active ingredient.

Published studies of residential applicator and/or post-application exposure to carbaryl on turf are non-existent. In fact, we are aware of no study, public or private, that examined both applicator and post-application exposure on turf using the same chemical. The hose-end applicator study in the Pesticide Handler's Exposure Database (3) was a mixer/loader/applicator study rather than the subjects using a Ready-To-Spray product. The limited post-application studies on turf have involved choreographed routines that may not mimic either the degree or intensity of post-application exposure of a resident (4, 5).

The study described here was designed to monitor the absorbed dose of carbaryl in an adult applicator residing at each home and all occupants of the home aged four years and older. The four-year age was selected as a cut-off for urine collection because of anticipated difficulties in reliable urine collection with younger children. At each home carbaryl was applied as a hose-end broadcast application to the lawn and to either a vegetable garden or ornamental flowers. Bayer CropScience selected the lawn and garden broadcast application of a liquid formulation of carbaryl because it represents the high-end use and is likely the highest exposure potential of carbaryl residential uses.

The post-application activities of the participants were not controlled in the study. The measurement of residential post-application exposure to pesticides has traditionally involved the use of a standardized activity such as the Jazzercise® routine. Study participants did keep a diary of their outdoor activities to assist in the interpretation of the exposure data. Therefore, the study was specifically designed to quantify the range of absorbed carbaryl doses in children and adults following an upper-bound, high-exposure, potential-use pattern. This range of actual absorbed doses is intended to be compared to the estimated absorbed doses of carbaryl calculated by EPA using a combination of carbaryl-specific data, data developed by the Outdoor Residential Exposure Task Force (ORETF), and EPA's Health Effects Division (HED) Residential Exposure Standard Operating Procedures. The comparison of the actual absorbed doses of carbaryl resulting from actual residential use and post-application activities provides a real-world context to interpret EPA's modelled estimate of residential exposure.

Overview of US EPA Screening-Level Risk Assessment

The Reregistration Branch 1 of US EPA's HED conducted a detailed residential exposure and risk assessment of carbaryl (6, 7). The EPA assessment addressed 17 potential residential use scenarios and quantified the potential exposure to a homeowner applying carbaryl around the home and the resultant post-application exposure potential. Three distinct subpopulations were evaluated for residential post-application exposure. These subpopulations were residential adults (aged 18 and up) youth-aged children (ages 10 to 12), and toddlers age three representing children between one and six years of age.

Three specific post-application scenarios were addressed in the EPA assessment of residential uses of carbaryl. The first scenario involved exposures to children playing on treated turf and adults on treated turf and specifically focused on young children as a sentinel group. The second scenario involved adults and youth-aged children working in the home garden and the third scenario involved the exposures to children following contact with pets treated with carbaryl. Because the biomonitoring study was not designed to address this latter scenario there is no further discussion of EPA's assessment of exposure to treated pets.

Handler Application Exposure

Handler exposures from carbaryl application to lawns and gardens were quantified using exposure data submitted either by Bayer CropScience or ORETF. ORETF Study OMA004 (MRID 449722-01) quantified the exposure to homeowners applying a pesticide to turf using either a dial-type (DTS) hose-end sprayer that requires the individual to pour concentrate formulation into the sprayer or a ready-to-spray (RTS) hose-end sprayer that contains the concentrated formulation in the sprayer at the time of purchase and therefore eliminates the need for the homeowner to pour concentrated formulation. Exposure to homeowners applying a 21% active ingredient liquid formulation of carbaryl by DTS hose-end sprayers or low pressure hand wand sprayers to vegetable gardens was quantified from a carbaryl study previously submitted to EPA by Bayer CropScience where dermal and inhalation exposure were measured by passive dosimetry (MRID 44459801).

The EPA assessment for lawn application (Scenario 8) was based only on the DTS hose-end sprayer. The dermal and inhalation exposures for a homeowner wearing shorts and a short-sleeved shirt were 11 mg/lb active ingredient (a.i.) and 0.016 mg/lb a.i., respectively. The dermal and inhalation exposure estimates for vegetable garden hose-end sprayer application were 34 mg/lb a.i. and 2 mg/lb a.i., respectively. A short-sleeve shirt and shorts scenario was used by EPA.

EPA's lawn application scenario assumed that five pounds of active ingredient were used to treat 20,000 square feet of lawn. This is representative of the current label maximum application rate of 10.9 pounds a.i. per acre (lb a.i./A). For perspective the amount handled of a standard 20% formulation would be approximately 12 one-quart containers of the ready-to-spray product. The resultant dermal and inhalation exposures calculated by EPA for the DTS hose-end sprayer were 0.786 mg/kg/day and 0.00114 mg/kg/day, respectively. Based on the dermal absorption value of 12.7% the absorbed dose is 0.101 mg/kg/day. The absorbed dose estimate permits comparison to the monitored absorbed carbaryl doses in the biomonitoring study.

The vegetable garden scenario for vegetables only was a subset of the Garden: Hose-End Scenario 3. The vegetable garden scenario assumed the use of one container to 1,000 square feet requiring the application of 0.012 lb a.i. (0.52 lb a.i./A). A second vegetable garden scenario assumed the use of one container requiring the application of 0.047 lb a.i. (2.0 lb a.i./A). The unit dermal and inhalation exposure estimates used for both scenarios were 34 mg/lb a.i. and 0.002 mg/lb a.i., respectively. The daily dermal and inhalation exposures resulting from treating the vegetable garden with 0.012 lb a.i. were 0.00583 mg/kg/day and 3.4×10^{-7} mg/kg/day, respectively. The absorbed dose for the 0.012 lb a.i. scenario is 0.00074 mg/kg/day. The daily dermal and inhalation exposures resulting from treating the vegetable garden with 0.047 lb a.i. were 0.0228 mg/kg/day and 1.3 x 10^{-6} mg/kg/day, respectively. This results in an absorbed dose of 0.0029 mg/kg/day.

Adult Post-Application Exposure

Adult post-application exposures relevant for comparison to the biomonitoring study were calculated by EPA for lawns and vegetable gardens. The assessment for lawns was conducted based on a high application rate of 8.17 lb a.i./A and a lower range application rate of 4.0 lb a.i./A. The turf transferable residues (TTRs) were obtained from the Georgia site of a carbaryl TTR study representing the highest measured value from the three sites tested in the US (MRID 451143-01). The Day 0 TTR was 1.20% of the application rate. Standard residential standard operating procedures (SOP) defaults (*8*) of two hour exposure to the lawn and a short-term/intermediate-term transfer coefficient (TC) of 14,500 cm^2/hr were used per EPA (*10*). Based on these data and defaults, the Day 0 adult dermal exposure to lawn carbaryl residues was 0.465 mg/kg/day at the high application rate and 0.228 mg/kg/day at the low application rate. The adult post-application absorbed dose estimates are 0.059 mg/kg/day at the 8.17 lb a.i./A application rate and 0.029 mg/kg/day at the 4.0 lb a.i./A application rate.

The post-application exposure to adults in vegetable gardens was based on carbaryl specific data generated by the Agricultural Reentry Task Force (ARTF). The dislodgeable foliar residues (DFRs) of carbaryl on vegetable crops were obtained from the ARTF Cabbage Weeding Study (MRID 451917-01). The Day 0 carbaryl DFR was 2.46 µg/cm^2 at the study application rate of 2.07 lb a.i./A. This was adjusted to the typical residential garden application rate of 2.0 lb a.i./A and resultant DFR of 2.38 µg/cm^2. The TCs used in the EPA assessment were 500 cm^2/hr for low exposure activities such as thinning or weeding young plants, 700 cm^2/hr for medium exposure activities such as scouting mature garden plants, and 1,000 cm^2/hr for high exposure activities such as harvesting vegetables, staking, or tying up plants in the garden. The TCs were from HED Policy 3.1 (*10*). The exposure duration was 0.67 hr/day from the Residential SOPs (*8*). Based on these data and defaults EPA estimated the daily dermal exposure to an adult on the day of application to be 0.011 mg/kg/day for low exposure activities, 0.016 mg/kg/day for medium exposure activities, and 0.023 mg/kg/day for high exposure activities. The resultant absorbed doses for low, medium, and high activities are 0.0014 mg/kg/day, 0.0020 mg/kg/day, and 0.0029 mg/kg/day, respectively.

Youth Post-Application Exposure

The post-application exposures relevant for comparison to the biomonitoring study of children aged 13-17 were calculated by EPA for 10-12 year old children working in vegetable gardens. The post-application exposure to older children in vegetable gardens was based on the same data as adults with age specific and body surface area specific adjustments to the TCs. The TCs used in the EPA assessment were 250 cm^2/hr for low exposure activities such as thinning or weeding young plants, 350 cm^2/hr for medium exposure activities such as scouting mature garden plants, and 500 cm^2/hr for high exposure activities such as harvesting vegetables, staking, or tying up plants in the garden. A body weight of 39.1 kg was used. The exposure duration was 0.67 hr/day from the Residential SOPs (8). Based on these data and defaults EPA estimated the daily dermal exposure to an older child on the day of application to be 0.010 mg/kg/day for low exposure activities, 0.014 mg/kg/day for medium exposure activities, and 0.020 mg/kg/day for high exposure activities. The resultant absorbed doses for low, medium, and high activities are 0.0013 mg/kg/day, 0.0018 mg/kg/day, and 0.0025 mg/kg/day, respectively.

Toddler Post-Application Exposure

EPA conducted a post-application exposure assessment for toddlers contacting carbaryl residues following turf and pet applications. Comparisons between the EPA assessment and the biomonitoring study are relevant for the turf assessment. Toddler post-application exposure to a liquid formulation lawn broadcast application involved dermal exposure and theoretical incidental oral ingestion via hand-to-mouth, object-to-mouth, and soil ingestion.

The assessment for lawns was conducted based on a high application rate of 8.17 lb a.i./A and a lower range application rate of 4.0 lb a.i./A. The turf transferable residues (TTRs) were obtained from the Georgia site of a carbaryl TTR study (MRID 451143-01). The Day 0 TTR was 1.20% of the application rate. Standard Residential SOP defaults (8) of two hour exposure to the lawn, a 15kg body weight, and a short-term/intermediate-term transfer coefficient (TC) of 5,200 cm^2/hr were used. Based on these data and defaults, the Day 0 toddler dermal exposure to lawn carbaryl residues was 0.778 mg/kg/day at the high application rate and 0.381 mg/kg/day at the low application rate. The absorbed dermal doses for a toddler are 0.099 mg/kg/day at the high application rate and 0.048 mg/kg/day at the lower application rate.

The toddler hand-to-mouth ingestion was based on a transfer rate from the turf to the hands of 5% of the application rate. A 20 cm^2 area of the hands was assumed to be inserted into the mouth and the contact occurs 20 times/hour with a two hour duration. For object-to-mouth ingestion, the transfer rate of carbaryl residues from the lawn to the object was assumed by EPA to be 20% of the application rate with a 25 cm^2 surface area of the object being mouthed. The saliva was assumed to extract 50% of the object's carbaryl residues. The soil ingestion estimation used by EPA assumed 100% deposition of the application rate into the top layer of the soil. The soil density was 0.67 cm^3/g and 100 mg of soil was assumed to be ingested daily. These incidental ingestion assumptions are consistent with EPA's Residential Exposure SOPs. At the high application rate of 8.17 lb a.i./A the hand-to-mouth, object-to-mouth, and soil ingestion Day 0 dose estimates were 0.122 mg/kg/day, 0.031 mg/kg/day, and 0.00041 mg/kg/day, respectively. The total incidental ingestion of carbaryl at the high application rate would be 0.153 mg/kg/day. At the lower application rate of 4.0 lb a.i./A the hand-to-mouth, object-to-mouth, and soil ingestion Day 0 dose estimates were 0.060 mg/kg/day, 0.015 mg/kg/day, and 0.00020 mg/kg/day, respectively. The total incidental ingestion of carbaryl at the 4.0 lb a.i./A application rate would be 0.075 mg/kg/day.

The EPA lawn and vegetable garden exposure estimates were based on a combination of product-specific monitoring and exposure data, surrogate exposure data, and default assumptions intended to provide upper-bound estimates of exposure. The resultant exposure estimates derived from these data and assumptions were then used by EPA to make risk management decisions. In the absence of data, such as obtained from the biomonitoring study, the representativeness of the estimates as typical or upper bound estimates of carbaryl dose is uncertain or unknown. The total application and post-application combined absorbed doses for lawn and vegetable gardens calculated by EPA are presented in Table I. The absorbed dermal dose was calculated from the dermal exposure by adjusting the dermal exposure by a 12.7% dermal absorption.

Overview of Carbaryl Residential Biomonitoring Study

The residential biomonitoring study was designed to characterize the range of absorbed carbaryl dose during and after the application of carbaryl to the lawn and either a vegetable garden or ornamental flowerbeds. Sevin GardenTech Ready-To-Spray is a common, commercially available carbaryl home product found in garden supply stores and hardware stores. All families participating in the study were provided with, and signed, consent forms. The families were recruited from pools of individuals intending to conduct home pesticide applications. The protocol was reviewed by an Independent Review Board (IRB) and complied with the Common Rule (Federal Policy for the Protection of Human Subjects of Research, 45 CFR Part 46) regarding studies involving

Table I. Summary of Lawn and Garden Exposures and Total Absorbed Doses Estimated by EPA

Population	Dose (mg/kg/day)
Adult	
DTS lawn application	0.101
Vegetable garden application	0.00074
Lawn re-entry (8.17 lb a.i./A)	0.059
Garden re-entry (high exposure)	0.0029
Combined Adult Absorbed Dose	0.164
Youth	
Garden re-entry (high exposure)	0.0025
Toddler – Lawn Re-Entry (8.17 lb a.i./A)	
Dermal	0.099
Hand-to-Mouth	0.122
Object-to-Mouth	0.031
Soil Ingestion	0.00041
Combined Toddler Absorbed Dose	0.252

human participants. The product contains 22.5% carbaryl as the active ingredient and is packaged as a RTS hose-end sprayer. The product is packaged with 32 fluid ounces of formulation in the sprayer.

Bayer CropScience (BCS) selected the lawn and garden application scenario because it would represent a high-end use of carbaryl by homeowners. The liquid formulation was chosen instead of the dust or granular formulations because the potential absorbed doses received from the liquid formulation would exceed that received from solid formulations. Liquid formulation exposures tend to be higher because more active ingredient is used in liquid applications and the concentration of active ingredient in liquids tends to be higher than in solid formulations. In addition, the dust formulation is not applied as a broadcast to the lawn. BCS has submitted an analysis of carbaryl residential use patterns based on the Residential Exposure Joint Venture's (REJV) National Pesticide Product Use Survey to EPA (5 June 2002, MRID 456905-01). The REJV survey is a 12-month diary survey in which participants recorded every application of each pesticide product around their homes. The REJV survey indicated that the majority of carbaryl applications involved the use of the 5% or 10% dust. In addition, vegetable gardens or ornamentals were the primary sites of application. Lawn application of the dust formulations were spot treatments and accounted for 6% of reported sites treated with the dust formulations. The use of concentrated formulations, such as the 22.5% a.i. formulation, accounted for 22% of inventoried household carbaryl products. Vegetable gardens, ornamental flowers, and shrubs accounted for 68% of all sites treated. Broadcast and spot lawn applications accounted for 7% of the sites treated. Although the broadcast lawn application requires the greatest amount of carbaryl for an application and provides a much greater opportunity for post-application contact, especially to young children than the vast majority of residential carbaryl use patterns, the results of the biomonitoring study represent a high use pattern that comprises less then 10% of the actual use patterns around the home.

The study biomonitored ten families in Missouri and 13 families in California. All families were required to have at least one child age four to 17 to qualify. The total number of study participants was 106 of which 23 applied carbaryl, 22 were spouses, six were other adult residents such as grandparents, 13 were older children (13-17 years old), and 42 were young children (4-12 years old). The average amount of GardenTech Ready-To-Spray applied in Missouri was 410 grams a.i. with a range of 332 to 613 g a.i. The lawn sizes in Missouri ranged from 4,095 ft^2 to 12,640 ft^2 and averaged 7,383 ft^2. The vegetable garden or ornamental area in Missouri ranged from 504 ft^2 to 5,800 ft^2 with an average of 1,595 ft^2. In California 255 g a.i. were applied to each lawn and garden area. The treated area averaged 3,196 ft^2 and ranged from 150 to 12,000 ft^2.

Adult Residential Application and Post-Application Doses

Complete urine voids were collected from all study participants beginning two days prior to application and continuing until Day 3 post-application. The two days of pre-application urine collections were important because 1-naphthol is not unique to carbaryl and may also indicate exposure to naphthalene, which can be found in cigarette smoke and solvents. The 1-naphthol levels found in the urine prior to application are considered baseline from carbaryl and other sources and the difference between the individual's baseline 1-naphthol levels and those observed after the application are indicative of the additional body burden of 1-naphthol that probably resulted from the carbaryl application.

The absorbed doses of carbaryl (corrected from the 1-naphthol) are presented in Table 7 of the biomonitoring report submitted to the US EPA (*11, 12*). All applicators in Missouri were male adults between the ages of 32 and 50. In California, five of the 13 applicators were females and the ages of all applicators ranged from 18 to 58. A summary of the pre-application absorbed dose levels for the applicators is presented in Table II. All dose levels are presented as carbaryl equivalents derived from the 1-naphthol biomarker for the basis of comparing to Day 0 through Day 3 dose levels.

Table II. Summary of Applicator Carbaryl Dose Levels (μg/kg) Prior to Application

Site and Day	Geometric Mean	Minimum	Maximum
MO Day -2	0.201	0.005	1.85
MO Day -1	0.420	0.005	5.8
CA Day -2	0.116	0.005	1.0
CA Day -1	0.103	0.005	1.48

1-naphthol levels in the adult U.S. population was measured by Hill, et al (13). In that study 1-naphthol was detected in 86% of 983 individuals monitored. The median 1-naphthol level detected was 4.4 μg/L. Assuming an adult daily urine volume of 1.5 L/day, a 70-kg body weight, and using the conversion value of 3.5 to equate 1-naphthol to carbaryl, the median carbaryl level in the U.S. population is 0.33 μg/kg assuming that all 1-naphthol is derived from carbaryl. The carbaryl biomonitoring study data are therefore consistent with those observed in the Hill reference population. Hill observed a 95th percentile value of 43 μg/L, which would be equivalent to 3.2 μg/kg using the above assumptions. Again, the Hill data are comparable to the maximum levels observed in the biomonitoring study prior to the carbaryl application.

The application of carbaryl by the homeowner lead to an increase in 1-naphthol excretion that was evident in the urine collected on the day of application (Day 0). The 1-naphthol urine concentrations dropped on Day 1 to Day 3 post-application in the Missouri applicators. This pattern is consistent with the pharmacokinetics of carbaryl that shows rapid excretion (14, 15, 16). The excretion pattern in the applicators was potentially confounded by the continued possible exposure to carbaryl residues from the treated lawn, vegetable garden, and flowers.

Table III provides a summary of the applicator absorbed carbaryl dose levels in Missouri and California for the application day and subsequent three days of post-application monitoring.

Table III. Summary of Applicator Carbaryl Geometric Mean Dose Levels[a] Days 0-3

Site	Day 0 (μg/kg)	Day 1 (μg/kg)	Day 2 (μg/kg)	Day 3 (μg/kg)
Missouri	8.30	6.10	2.58	1.94
California	1.65	1.58	1.14	1.80

[a] Not adjusted for background 1-naphthol levels.

Because the study was set up to evaluate an upper-bound use pattern (lawn broadcast hose-end spray and concurrent vegetable garden hose-end spray), the resultant dose levels represent high-end exposures. Other typical use patterns and formulations of carbaryl with exposures that are expected to be lower include lawn spot treatment, dust applications to gardens and flowers, perimeter applications, pet collars, and granular lawn broadcast applications. Because the study involved RTS hose-end sprayers, the only major use pattern possibly not covered by the biological monitoring data would be the DTS hose-end lawn broadcast application. Since the DTS hose-end application is so much more labor intensive, it is unlikely that a person would use nearly as much by DTS as RTS.

The applicator-absorbed doses can be compared to the ORETF homeowner hose-end sprayer data reported in Study OMA004 (unpublished data). The RTS hose-end sprayers produced a dermal exposure with shorts and t-shirt of 2.3 mg/lb a.i. and an inhalation exposure of 0.010 mg/lb a.i. Adjusting the dermal exposure for the 12.7% dermal absorption produces an absorbed dermal dose of 0.292 mg/lb a.i. The average amounts of carbaryl sprayed in Missouri and California were 0.90 lb a.i. and 0.56 lb a.i., respectively (410 g a.i. and 255 g a.i.). Based on the ORETF data, the amount of a.i. applied, and a 70-kg body weight yields predicted absorbed doses of 3.8 µg/kg in Missouri and 2.3 µg/kg in California, and are quite consistent with biomonitoring measurements reported in this study. The RTS hose-end sprayer exposures were not calculated by EPA. The dermal and inhalation unit exposures from the ORETF study were 2.3 mg/lb a.i. and 0.010 mg/lb a.i., respectively. Based on the 12.7% dermal absorption the combined absorbed dose is 0.30 mg/lb a.i. The absorbed dose for a 70-kg person handling 1 lb a.i. is 4.3 µg/kg/day. The ORETF-based estimate does not include any post-application exposure.

The EPA-predicted exposures for an adult treating the lawn, garden, and having post-application garden exposure was converted to total absorbed dose estimates in Table I for comparison with the biomonitoring data. The comparison cannot be exact because the application rates used by the applicators varied, clothing varied, area treated varied, and the post-application activities varied. However, the summary is insightful in comparing and interpreting a standard deterministic exposure assessment using label application rates, SOP transfer coefficients, and default post-application activity pattern assumptions (see Table IV). The EPA absorbed dose predicted for an adult using a DTS hose-end sprayer to conduct a lawn broadcast application at 10.9 lb a.i./A, a concurrent garden application, and re-entry to the lawn and garden on the day of application would be 164 µg/kg/day for Day 0 alone. This value is approximately nine-fold higher than the mean total absorbed dose over four days observed in Missouri where the applicators handled 0.90 lb a.i. compared to the EPA assumption of 5.0 lb a.i. Assuming direct linearity between application rate and absorbed dose would explain much of the difference between the EPA predicted absorbed dose and the actual absorbed dose. Adjusting the 164 µg/kg/day predicted absorbed dose for the 5.5-fold difference in carbaryl handled between the EPA scenario and the Missouri scenario reduces the EPA based absorbed aggregate adult dose to 33 µg/kg/day.

The total four-day combined California applicator mean absorbed dose was 6.2 µg/kg/day. Similar adjustments in the amount of active ingredient handled can also be done to permit a more direct comparison. All California applicators handled 0.56 lb a.i. or 8.9 times less than the EPA scenario used for comparison. The adjusted EPA aggregate absorbed dose would be 18 µg/kg/day compared to the total California absorbed applicator dose of 6.2 µg/kg/day. The similarity of the absorbed doses would again be even closer if EPA had used the RTS hose-end sprayer data.

Table IV provides a comparison of the Missouri and California biological monitoring absorbed doses to the EPA-based estimate and the ORETF ready-to-spray (RTS) sprayer estimate when all estimates are normalized to 1.0 lb a.i.

Table IV suggests that when normalized to pounds handled, EPA's assessment of adult applicator and post-application exposure is reasonable. It is EPA's assumption that 5.0 lb a.i./day are sprayed by the consumer that is not in agreement with the actual use patterns observed in Missouri and California and obtained from the REJV survey.

Adult Non-Applicator Post-Application Doses

An evaluation of the absorbed dose levels of the adults not involved in the application of carbaryl provides important information regarding secondary routes of exposure such as track-in or contact with pets, clothing, or other objects that may have been in contact with treated surfaces after application. This population provides such information because they were not involved in the carbaryl application and often did not have direct post-application contact with the treated lawn or gardens.

All Missouri adults not involved in application were designated as spouses in the report and were females between the ages of 32 and 50. The average age was 39. By study definition all of the Missouri spouses had at least one child age 17 or younger living in the house. The California adults monitored in the study who did not apply carbaryl were more diverse than in Missouri. The California families appeared to include grandparents, adult children, and non-traditional arrangements. For example, California Site 8 involved a single parent family consisting of a 58 year old female head of household (considered a resident) and a five year old female child. The 12 California residences monitored with two or more adults, where the third adult (and any other

Table IV. Summary of Biomonitoring[a] vs. Passive Dosimetry[b] Exposure Estimates

Site	Amount Handled (lb)	Absorbed Dosage (µg/kg)	Normalized Dose (µg/kg/lb ai)
Missouri	0.9	18.9	21
California	0.56	6.2	11
EPA RED[c]	5.0	164	33
ORETF RTS	1.0	4.3	4.3

[a] Missouri and California
[b] EPA RED and ORETF RTS
[c] Reference *17*.

additional adults) was considered a resident, involved seven non-applicator adult males ranging in age from 18 to 73 years with an average age of 38. There were 11 adult females monitored and they ranged in age from 18 to 67. The average female adult non-applicator age was 37.

Table V provides a summary of the geometric mean carbaryl doses of the adults in Missouri and California. The non-spousal adults in California are referred to as residents. It is important to understand that not all of the pre-application 1-naphthol detected in the urine was present as a result of exposure to carbaryl. Because it is not possible to segregate the 1-naphthol between carbaryl and non-carbaryl sources the report has treated all 1-naphthol as carbaryl derived.

Table V. Summary of Adult Non-Applicator Geometric Mean Carbaryl Dose Levels (µg/kg)

Site	Day -2	Day -1	Day 0	Day 1	Day 2	Day 3
Missouri spouses	0.42	0.093	0.57	0.77	0.92	1.85
CA spouses	0.119	0.214	0.54	0.96	0.83	0.423
CA residents	0.07	0.252	2.27	1.32	0.92	2.89

The pre-application background levels among the individual adults ranged from non-detectable levels to 5.8 µg/kg. Although there is an obvious trend toward slightly higher carbaryl levels after application, the mean post-application values are all within the range of background levels of carbaryl based on 1-naphthol. The highest post-application absorbed dose observed in Missouri was 4.9 µg/kg, which indicates that the magnitude of post-application levels was similar to the pre-application levels. In California, all post-application dose levels were also below the 5.8 μg/kg pre-application dose levels with the exception of two individuals. The spouse at California Site 15 had a carbaryl dose level of 8.2 μg/kg on Day 1 and a 67 year old female at California Site 5 had dose levels of 11.4 μg/kg, 12.2 μg/kg, and 12.6 μg/kg on Days 1, 2, and 3, respectively. Based on the activities recorded in Table 3 of the study report (12), the spouse had no obvious contact with treated surfaces while the 67 year old resident worked on the treated lawn for two hours per day on Days 1, 2, and 3 post-application.

Post-Application Exposure to Children 13-17 Years Old

The absorbed 1-naphthol doses of the 13 children, aged 13 to 17, provide insight into the importance of post-application activity on exposure potential. A summary of the geometric mean absorbed doses is provided in Table VI. The exposure pattern and magnitude of exposure for this sub-population is similar to that observed for adults not involved in the application of carbaryl.

Table VI. Summary of 13-17 Year Old Children Geometric Mean Carbaryl Dose Levels (μg/kg)

Site	Day -2	Day -1	Day 0	Day 1	Day 2	Day 3
Missouri	0.035	0.083	1.46	2.17	0.89	1.49
California	0.146	0.036	0.490	2.20	5.85	5.60

The pre-application carbaryl doses among the older children (Day -1 and Day -2) ranged from non-detectable levels to 0.14 μg/kg assuming all 1-naphthol is derived from carbaryl. The upper end of the range was lower than the background observed in adults that ranged up to 5.8 μg/kg (Table II). Among the 13 children in this cohort the carbaryl application did produce an apparent increase in the carbaryl dose. In most children the increase was very small. Five of the 13 children had carbaryl dose levels that were less than twice the background maximum of 1.4 μg/kg. Four of those children were from Missouri (Sites 5, 7, 8, and 9) and three had minimal post-application activity on the lawn. The child at Site 5 recorded less than half of an hour on Day 2 and Day 3, the child at Site 7 had no recorded yard activity, and the child at Site 9 played in the yard for two hours on Day 1. Site 6 and that child had no recorded yard activities. However, the 16 year old child at Missouri Site 8 reported one to six hours of yard activity during Day 0 to Day 3 post-application. The intensity of the activities was not reported. The carbaryl dose levels increased to 0.729 µg/kg/day to 2.4 µg/kg/day. The increase in the 16 year old Missouri Site 8 dose levels appeared large because the child had non-detectable levels during the pre-application period. Although the intra-personal increase in dose level was large, the magnitude of the carbaryl exposure was comparable to the general background levels.

Larger increases in carbaryl dose levels were observed in three of the seven older Missouri children. In each case some outdoor activity was recorded. All of the measured levels were relatively low with the highest recorded carbaryl dose among the older Missouri children being 12.6 µg/kg at Missouri Site 2. The child at Site 2 was 13-years old and recorded activity in the yard on the day of application and each day thereafter. The Day 0 and Day 1 carbaryl levels increased from the background dose level of 1.24 µg/kg on Day -1 to 12.6 µg/kg on Day 0 and 10.9 µg/kg on Day 1. The dose levels dropped off to background levels of 0.763 µg/kg and 1.62 µg/kg on Day 2 and Day 3 post-application, respectively, confirming the rapid clearance following exposure. The 13 year old at Missouri Site 6 recorded significant amounts of time in the yard at 2.75 hours on the day of application, 1 hour on Day 1, 2.5 hours on Day 2, and 45 minutes on Day 3. Despite the significant duration of outdoor activity the carbaryl exposure on Day 0 through Day 3 post-application never exceeded a four-fold increase above the child's Day -1 dose level of 0.968 µg/kg. The highest dose level recorded for the child was 3.91 µg/kg on Day 1. The 15 year old child at Missouri Site 11 watched the carbaryl application for 30 minutes. The child's carbaryl level increased from non-detectable on the two days prior to application to 7.99 µg/kg on Day 0 when she watched the application. The dose levels dropped on Day 1 to 4.18 µg/kg and 0.75 µg/kg on Day 2 post-application. This child's carbaryl level increased to 9.15 µg/kg on Day 3. No yard activity was recorded for her to explain the increase. The younger siblings of the 15 year old girl had increased carbaryl levels on Day 3 compared to Day 2. The 12 and nine year old siblings also reported no yard activity while the five and four year olds did record 30 minutes of activity on Day 3. On Day 2 the four younger siblings all reported 30 minutes of yard activity. The Day 3 increase for the 15 year old girl is not readily explained by the available activity information. However, the increase in all children at Missouri Site 11 on Day 3 could indicate a potential exposure to either a non-carbaryl source of 1-napthol or exposure to carbaryl unrelated to direct contact with lawn residues.

The carbaryl dose levels observed among the California children ages 13 to 17 provide an insight into the interaction of the application rate and activities. Unlike Missouri the application rates in California varied greatly. Two of the six residences in California that had older children also had application rates comparable to the Missouri sites. California Site 7 had an application rate of 2.7 lb a.i./A and a 13 year old with 1.5 hours of yard activity on the day of application. The child's carbaryl levels increased but remained near the overall background levels with a range of 0.807 µg/kg to 3.95 µg/kg. A 14 year old at California Site 5 also lived at a residence with an application rate of 2.7 lb a.i./A that is comparable to the Missouri application rates. The child had post-application yard activities on Day 1 and Day 2 of two hours each day. The carbaryl exposure levels were in the range observed in Missouri and ranged from 3.66 µg/kg to 8.21 µg/kg following the yard activity. The Day 0 carbaryl level was 1.21 µg/kg and was within the background dose level range.

The remaining four California sites with older children had application rates beyond the recommended label lawn rate. When no yard activity was recorded the carbaryl dose levels were not significantly increased. The child at California Site 6 had dose levels remaining in the background range as previously discussed. California Site 15 had the highest application rate. The 14 year old child at this site reported no yard activities following application. The child's carbaryl dose level (based on 1-naphthol) increased from 1.4 µg/kg prior to the carbaryl application to 5.2 µg/kg after application. The results at this site suggest that regardless of how high the application rate there is minimal effect on body levels of carbaryl when direct contact with the treated area is avoided. This strongly suggests that secondary exposures have little impact on absorbed doses in older children.

Two older children had both post-application yard activities and high application rates to their lawns. At California Site 3 the child's carbaryl levels were 1.07 µg/kg on the day of application but rose to 34.3 µg/kg on Day 1 and then declined to 28.7 µg/kg on Day 2 and 14.8 µg/kg on Day 3 post-application. Activity in the yard of unspecified duration occurred on the day of application followed by one hour on Day 1. No entries were made in the activity diary for the subsequent two days. The 13 year old at California Site 11 had no activity on the day of application or Day 1 and had dose levels of 0.442 µg/kg on Day 0 and 1.06 µg/kg on Day 1. One hour of activity in the yard occurred on Day 2 and the carbaryl levels increased to 57.5 µg/kg with a decline to 22.9 µg/kg on Day 3, which also involved no yard activity.

Based on the results of the older child cohort it becomes evident that this group is substantially similar to adults not involved in applying. Only when post-application activities occurred at California homes treated at higher application rates did significant increases in carbaryl dose levels occur.

Post-Application Exposure to Children 4-12 Years Old

The background levels among the Missouri children ranged from 0.005 µg/kg to 12.5 µg/kg and among the California children the range was 0.005 to 2.21 µg/kg. Table VII shows a summary of daily geometric mean results by day for the younger children in this residential monitoring study. This group (n = 42) constituted the largest subset of the entire study. 1-naphthol levels among children in Minnesota were measured by Adgate, et al (*18*). The study was part of the National Human Exposure Assessment Survey (NHEXAS) and evaluated 102 children age three to 13 in Minnesota. The geometric mean 1-naphthol level detected was 1.4 µg/L. The average daily urine volume of the Missouri four to 12 year olds was 594 mL/day and the average for the California children was 501 mL/day. Assuming a daily urine volume of 0.55 liters/day, a 28-kg body weight (the average of the 42 children age four to 12), and using the conversion value of 3.5 to equate 1-naphthol to carbaryl, the average background carbaryl level among the Minnesota children is 0.096 µg/kg assuming that all 1-naphthol is derived from carbaryl. The Minnesota children's geometric mean levels were within the range observed in the carbaryl biomonitoring study prior to application. The Minnesota study observed a maximum value of 55 µg/L, which would be equivalent to 3.8 µg/kg using the above assumptions. The Minnesota maximum value is at the 97th percentile of the distribution of the Missouri and California children's background levels. The 95th percentile of the background levels among the California and Missouri children was 2.64 µg/kg.

Table VII. Summary of 4-12 Year Old Children Geometric Mean Carbaryl Dose Levels (µg/kg)

Site	Day -2	Day -1	Day 0	Day 1	Day 2	Day 3
Missouri	0.133	0.0403	1.96	3.26	1.50	3.32
California	0.0419	0.0826	3.19	6.10	7.98	5.10

Maximum exposures in children age four to 12 are the highest measured in any post-application group. The younger children had consistently larger mean post-application exposures than the older children with the exception of the last study day in California. Exposures clearly increase following application in most, but not all children age four to 12. The increase appears to be the result of younger children spending more time in the treated area than older children in general. Although it is difficult to associate exposure with any particular activity or duration of activity, it is possible to relate increased 1-naphthol excretion with spending any time in the yard following treatment. Because of the size of the subset, a comparison between siblings that did and did not report yard activity was possible.

Four children in Missouri had exposures that did not increase over background. Both children at Site 3, the seven year old at Site 1, and the eight year old at Site 9 had post-application urine levels that did not exceed the 95th percentile of the pre-application levels. Four children in California had post-application urine levels that never exceeded the 95th percentile of the pre-application levels. With the exception of the eight year old at Missouri Site 9 all of these children had no reported contact with the treated lawn. Three of the four California children lived at households where the application rate was high. These results further reinforce the observation that application rate is not the primary determinant of exposure and that direct contact with treated surfaces is a primary determinant of exposure. The fact that the children who did not contact the treated lawns with higher application rates did not show increases above background, also shows that secondary routes of exposure such as vapor incursion, track-in, and dust contact played little or no role in the children's absorbed doses.

Among the Missouri children age four to 12 that had contact with the treated yard, eight had absorbed dose levels that exceeded 10 μg/kg. Four of these children resided at Missouri Site 9. The Day -2 pre-application 1-naphthol levels were high among these children suggesting an additional source of 1-naphthol was present in addition to the carbaryl application. The 12 year old had a pre-application 1-naphthol level that converts to a carbaryl equivalency of 6.87 μg/kg, the ten year old a level of 12.5 μg/kg, and the four year old a level of 4.75 μg/kg. The parents, a 15 year old sibling, and the 6 and eight year old did not have elevated levels of 1-naphthol prior to application. All family members had very low to non-detectable levels on Day -1. This type of pattern within one family suggests that an alternative source of exposure to a 1-naphthol-producing chemical was occurring outside the home. The four children at Site 9 reported two hours of activity in the yard on Day 1 post-application, and this was concurrent with an increase in 1-naphthol levels between Day 0 and Day 1. The maximum carbaryl dose level peaked at 24.9 μg/kg on Day 1 for the ten year old. Interestingly, the eight year old at Site 9 was one of the four Missouri children whose dose levels remained within the background range. However, this child did show a slight increase in carbaryl compared to her pre-application levels.

The remaining three Missouri children that had post-application dose levels greater than 10 μg/kg all reported conducting activities in the yard following the carbaryl application. A seven year old at Missouri Site 2 had post-application carbaryl levels between 4.2 and 16.5 μg/kg and yard activities between one and 2.5 hours/day on each of the post-application days. The five year old at Missouri Site 6 had dose levels up to 13.1 μg/kg following yard activities that ranged from two to 3.5 hours on three of the four days following application. The ten year old at Missouri Site 8 had the highest carbaryl levels in Missouri. The child's dose level was 61.2 μg/kg on Day 0 and 55.8 μg/kg on Day 3. In between these days, the dose levels were less than 3 μg/kg, again suggesting very rapid clearance. The child also had reported yard activity durations of up to six hours per day. The two younger siblings of this child reported similar yard activity durations; however, their dose levels never exceeded 10 μg/kg. Because the specific yard activities and the intensity of the contact was not recorded to minimize the burden of study participation, a more detailed comparison of activity type with absorbed dose between the Site 8 siblings is not possible. The final Missouri child with a carbaryl level that exceeded 10 μg/kg was the four year old at Site 10. The child had a maximum level of 11.4 μg/kg on Day 3 following one hour of yard activity on Day 2 and 1.5 hours on Day 3. This child also had relatively high pre-application background levels in excess of 1 μg/kg.

Among the two California sites in which the lawn was treated at lower to average label rates there were no measured carbaryl levels that exceeded 10 μg/kg. The nine year old at California Site 1 had a maximum level of 7.62 μg/kg with reported activities of two hours in the yard treated at 8.1 lb a.i./A. Similarly, the five year old at California Site 8 did not exceed a dose level of 4.72 μg/kg with reported activity of one hour per day on three days with the lawn treated at 2.0 lb a.i./A. A total of eight of the 19 California children had no dose levels in excess of 10 μg/kg even when sites with the highest application rates were included.

When the combination of the higher application rates and yard activity occurred concurrently in California there were clear elevations of carbaryl dose levels. The results at these sites were extremely useful in understanding some of the dynamics involved in children's exposures to lawn and garden pesticide residues.

The most interesting dynamics occurred at California Site 11. Three children resided at this site and their yard activity durations differed. The eight year old had the highest carbaryl levels observed in the biomonitoring study at 201 μg/kg, 447 μg/kg, and 347 μg/kg on Day 1, 2, and 3 post-application, respectively. (For comparative purposes, the next highest level among any of the children was 77.6 μg/kg for the four year old at Site 13.) This child reported the most yard activities at 0.5 hours on Day 0 and 1.5 hours on Day 2, the day that the 447 μg/kg dose level occurred. Contrasting to the eight year old are the 11 and four-year old siblings. The 11 year old was reported to have 0.5 hours of yard activity on the day of application and no further contact with treated turf. Although the carbaryl dose level for the 11 year old was only 3.01 μg/kg on Day 0, it spiked at 44.9 μg/kg on Day 1 and rapidly declined to 10.8 μg/kg by Day 2 and 9.7 μg/kg on Day 3. The apparent lag between exposure on Day 0 and highest concentrations could have resulted from the contact occurring late on the day of application. The four year old had no reported yard contact and carbaryl dose levels in this child were the lowest of the three siblings. The carbaryl dose levels for the four year old ranged from 3.13 μg/kg to 17 μg/kg following the carbaryl application.

Another example of how activity on the lawn is the primary determinant for the resultant carbaryl dose levels is evident among the children at California Site 13 where the homeowner applied a higher rate of carbaryl. As previously discussed the eight year old and seven year old in this household had carbaryl dose levels that did not exceed 10 μg/kg (highest level was 6.4 μg/kg). Both children only had reported yard activities on Day 3. These results can be compared to the four year old sibling at California Site 13 who had reported one hour per day yard activities on the day of application and each day thereafter. This child's carbaryl levels were 2.34 μg/kg on Day 0 and rose as high as 77.6 μg/kg on Day 2. The combination of the repeated yard activity combined with an excessive application rate resulted in this child having the second highest carbaryl levels of all children in the study.

Conclusions

Bayer CropScience conducted a biological monitoring study of individuals residing in homes in California or Missouri in which carbaryl was applied by a member of the family. The study was designed to monitor the absorbed dose of carbaryl in an adult applicator residing in each of the homes and all occupants of the home aged four years or older. The study did not control the post-application activities of the study participants and provides a comparison to residential exposure estimates based on standardized routines such as the Jazzercise® choreographed activity pattern often used in re-entry exposure monitoring studies (1, 2, 4).

Exposure to applicators was consistent with previous observations when normalized for quantity of pesticide applied. Applicators who applied more carbaryl generally had increased exposure. For this reason, Missouri applicators that handled up to twice as much carbaryl had proportionately higher exposures relative to their California counterparts. Adult post-application exposure was very low and difficult to discern from background levels. This pattern was even evident in California where nine of 13 Sites were treated at the highest application rates. Older children age 13-17 had exposures very similar to adult spouses or residents not involved in the application. This exposure pattern suggests that older children have very little post-application contact with treated lawns or gardens. Children in the 4-12 age group were the most highly exposed and the exposure can be linked to time spent in the treated area. Like most adults not involved in the application, the minority of children that did not go into the treated areas had concomitant exposures consistent with background levels. The determination that post-application dose levels among individuals of all age groups not reporting direct contact with the treated surfaces suggests that "secondary" exposure pathways (vapor incursion, track-in, and dust) are inconsequential.

Among children there was no apparent relationship between age and exposure. The youngest sibling was rarely the most highly exposed. Because hand-to-mouth activity decreases with age it would be expected that the youngest children would be the most highly exposed if mouthing activities were a significant route.

The most significant predictor of increased exposure was spending time on or in the treated area. Casual contact or indirect contact does not appear to yield significant exposure. Because of the significance of activity on treated surfaces the use of standardized routines to estimate exposure significantly overestimate the likely range of post-application exposures. This results from the design of the Jazzercise®-type routines that are intended to maximize activity contact with treated surfaces. EPA estimated toddler absorbed dose estimate of 252 µg/kg is 32 times greater than the highest daily mean absorbed dose measured among the 4-12 age group in this study. The highest individual dose level observed in Missouri where the application rates were comparable to the EPA assessment was 61 µg/kg. Only one child in California had dose levels similar to the EPA estimate. This child had dose levels up to 447 µg/kg that resulted from the combination of a higher application rate and direct contact activities in the treated yard.

References

1. Ross, J., Thongsinthusak, T., Fong, H.R., Margetich, S. and Krieger, R. 1990. Measuring Potential Dermal Transfer of Surface Pesticide Residue Generated from Indoor Fogger Use: An Interim Report. Chemosphere, Vol.20, Nos. 3/4:349-360.
2. Ross, J.H., Fong, H.R., Thongsinthusak, T., Margetich, S. and Krieger, R. 1991. Measuring Potential Dermal Transfer of Surface Pesticide Residue Generated from Indoor Fogger Use: Using the CDFA Roller Method. Interim Report II. Chemosphere, Vol.22. Nos.9-10:975-984.
3. Keigwin, T.L. 1998. PHED Surrogate Exposure Guide: Estimates of Worker Exposure From the Pesticide Handler Exposure Database Version 1.1. EPA, Office of Pesticide Programs, Washington, D.C.
4. Williams, R.L., Oliver, M.R. Ries, S.B.and Krieger, R.I. 2003. Transferable chlorpyrifos residue from turf grass and an empirical transfer coefficient for human exposure assessments. Bull. Environ. Contam. Toxicol. 70: 644-651.
5. Harris, S.A. and Solomon, K.R. 1992. Human exposure to 2,4-D following controlled activities on recently sprayed turf. J. Environ. Sci. Health B27: 9-22.

6. U.S. EPA (United States Environmental Protection Agency). 2002. Carbaryl: Occupational and Residential Exposure Assessment and Recommendations for the Reregistration Eligibility Decision Document. Memorandum of 29 May 2002 from Jeffrey Dawson to Anthony Britton. USEPA, HED, Reregistration Branch 1.

7. U.S. EPA (United States Environmental Protection Agency). 2003. Carbaryl: Revised HED Risk Assessment – Phase 5 – Public Comment Period, Error Correction Comments Incorporated; DP Barcode D287532, PC Code: 056801. March 14, 2003. (J.L. Dawson).

8. U.S. EPA (United States Environmental Protection Agency). 1997. Standard Operating Procedures for Residential Exposure Assessments. Prepared by the Residential Exposure Assessment Work Group. U.S. EPA, Office of Pesticide Programs and Versar, Inc. Contract No. 68-W6-0030.

9. U.S. EPA (United States Environmental Protection Agency). 2001. Science Advisory Council for Exposure Policy Number 12: Recommended Revisions to the Standard Operating Procedures (SOPs) for Residential Exposure Assessments. Revised February 22, 2001. U.S. Environmental Protection Agency, Office of Pesticide Programs, Washington, DC.

10. U.S. EPA (United States Environmental Protection Agency). 2000. Policy 3.10- Agricultural Transfer Coefficients, Revised August 7. Office of Pesticide Programs, Science Advisory Council for Exposure, Washington, DC.

11. Rice, F. 2002. Measurement of pesticide exposure of suburban residents associated with the residential use of carbaryl. ABC Laboratories, Inc., Columbia, MO. ABC Laboratories Study Number: 46335. October 8, 2002. MRID #45788501.

12. Rice, F. 2002. Amended Final Report - Measurement of pesticide exposure of suburban residents associated with the residential use of carbaryl. ABC Laboratories, Inc., Columbia, MO. ABC Laboratories Study Number: 46335. March 5, 2002. MRID #45897401.

13. Hill, RH, et al. 1995. Pesticide Residues in Urine of Adults Living in the United States: Reference Range Concentrations. Environmental Research 71, 99-108.

14. Krolski, M. E., Nguyen, T., Lopez, R., Ying, S.-L., and Roensch, W. (2004a). Metabolism and Pharmacokinetics of [14C] Carbaryl in Rats. Bayer Report 201025. MRID 46277001.

15. Krolski, M. E., Nguyen, T., Lopez, R., Ying, S.-L., and Roensch, W. (2004b). Metabolism and Pharmacokinetics of [14C] Carbaryl in Rats Following Mixed Oral and Dermal Exposure. Bayer Report 201026. MRID 46277002.

16. Ross, J.H. and Driver, J.H. (2002). Carbaryl mammalian metabolism and pharmacokinetics. Infoscientific.com, Inc. document prepared for Bayer CropSciences. MRID# 45788502.

17. U.S. EPA (United States Environmental Protection Agency). 2003. Interim Registration Eligibility Decision for Carbaryl (List A, Case 0080). June 30, 2003.   U.S. Environmental Protection Agency, Office of Pesticide Programs, Washington, D.C.

18. Adgate, J.L., D.B. Barr, C.A. Clayton, L.E. Eberly, N.C.G. Freeman, P.L. Lioy, L.L. Needham, E.D. Pellizzari, J.L. Quackenboss, A. Roy, and K. Sexton. 2001. Measurement of Children's Exposure to Pesticides: Analysis of Urinary Metabolite Levels in a Probability-Based Sample. Environmental Health Perspectives. Vol. 109, No. 6, pp. 583-590.

Chapter 15

Probabilistic Methods for the Evaluation of Potential Aggregate Exposures Associated with Agricultural and Consumer Uses of Pesticides: A Case Study Based on Carbaryl

Jeffrey H. Driver[1], John H. Ross[2], M. Pandian[3], Curt Lunchick[4], J. Lantz[4], G. Mihlan[4], and B. Young[4]

[1]infoscientific.com, Inc., 10009 Wisakon Trail, Manassas, VA 20111
[2]infoscientific.com, Inc., 5233 Marimoore Way, Carmichael, CA 95608
[3]infoscientific.com, Inc., 2275 Corporate Circle, Suite 220, Henderson, NV 89074
[4]Bayer CropScience, 2 T.W. Alexander Drive, Research Triangle Park, NC 27709

This chapter presents a case study illustrating probabilistic methods that can be used in analyses of potential aggregate daily exposures to pesticides. The methodological approach is illustrated for the agricultural and consumer product uses of carbaryl in the United States. Carbaryl is a broad-spectrum insecticide and has been used in U.S. agriculture and for professional turf management, professional ornamental production, and in various consumer products for ornamentals, vegetable gardens, fruit and nut trees, lawns, and pets. This carbaryl aggregate assessment case study presents methods that can be used to estimate potential daily dietary and non-dietary exposures to adults and children. Probabilistic aggregate exposure analyses provide a basis for investigating contributions from sources, considering variability and uncertainty, and can also provide a more rigorous basis for informed safety determinations.

Introduction

Assessment methods for addressing potential multi-source, multi-route aggregate exposures to pesticides have been developed and refined following promulgation of the U.S. Food Quality Protection Act (FQPA) of 1996 (http://www.epa.gov/opppsps1/fqpa/). FQPA amended the U.S. Federal Insecticide, Fungicide, and Rodenticide Act (FIFRA) and the U.S. Federal Food Drug, and Cosmetic Act (FFDCA). These amendments fundamentally changed the manner in which the U.S. Environmental Protection Agency (EPA) regulates pesticides. The requirements included a new safety standard, i.e., "reasonable certainty of no harm," that must be applied to all pesticides used on foods. This safety standard has resulted in the need to conduct quantitative human health risk analyses that include consideration of potential aggregate and cumulative exposures. Aggregate exposure represents exposure to individuals via all potential sources (food, water, and non-dietary, i.e., residential-related). Cumulative exposure represents the combined multi-source or route exposures from each chemical in a group of chemicals sharing a presumptive common mechanism of toxicity (http://www.epa.gov/pesticides/cumulative/).

To illustrate the methods for estimating aggregate pesticide exposures, a case study is presented that focuses on the insecticide carbaryl. Carbaryl is used in agriculture to control pests on terrestrial food crops including fruit and nut trees (e.g., apples, pears, almonds, walnuts, and citrus), many types of fruits and vegetables (e.g., cucumbers, tomatoes, lettuce, blackberries, and grapes), and grain crops (e.g., corn, rice, sorghum, and wheat). In the case of consumer products, carbaryl is used primarily for residential ornamental and garden care. Lawn, tree and pet care consumer product uses also exist, but their use is not as prevalent (particularly in the case of pet care products). Carbaryl consumer product formulations include dusts, ready-to-use sprayers containing liquid solutions, liquid concentrates, and granulars. Carbaryl outdoor consumer products are used for control of nuisance and economic pests (e.g., cutworms, crickets, white grubs), public health/disease vectors (e.g., ticks, fleas, mosquitoes), and imported fire ants. Carbaryl can be used by homeowners for treatment of ornamentals, vegetable gardens, residential turf (e.g., spot treatment of fire ant mounds), and on companion animals, albeit infrequently. There are no labels for indoor uses such as crack-and-crevice treatments of a residence. Homeowners can use a variety of application methods including ready-to-use (RTU) trigger sprayers, hose-end sprayers, and RTU dust packaging. Key physicochemical properties of carbaryl (naphthyl-methylcarbamate or 1-naphthyl methylcarbamate; CAS No. 63-25-2; EPA PC Code 56801) are summarized below in Table I.

Table I. Physicochemical Properties Carbaryl

Empirical Formula	$C_{12}H_{11}NO_2$
Molecular Weight	201.2
Appearance	White to tan solid
K_{oc}	217
Vapor Pressure	< 0.005 mm Hg at 26° C
Specific Gravity	1.23 at 20° C

A stochastic (or probabilistic), population-based (per capita), aggregate exposure assessment was developed for carbaryl using the CARES® (Cumulative Aggregate Risk Evaluation System) model. The CARES software and associated documentation is available from the International Life Sciences Institute's (ILSI) Research Foundation (http://cares.ilsi.org/). CARES represents a software program designed to conduct complex multi-source exposure and risk analyses for pesticides and other chemicals, such as the assessments required under the 1996 FQPA (http://www.epa.gov/opppsps1/fqpa/). CARES was originally developed under the auspices of CropLife America (CLA; http://www.croplifeamerica.org/), which conceived the project, provided funding, and managed the program's evolution. Scientific and technical contributions to the program's development came from a broad team of experts, including scientists from CLA's member companies and staff, consulting companies, EPA, and the U.S. Department of Agriculture (USDA). With its transfer to the ILSI Research Foundation, the CARES program continues to be publicly available at no charge. Also, stakeholders and interested parties continue to be invited to contribute to the technical and scientific advancement of CARES.

The CARES aggregate assessment for carbaryl, similar to many assessments for pesticides in the U.S., is based on currently available data, i.e., publicly available data such as EPA's Food Consumption Intake Database (FCID), the USDA's Pesticide Data Program (http://www.ams.usda.gov/science/pdp/), and proprietary chemical-specific data developed by carbaryl registrant's such as food, water and residential environmental residue monitoring studies sponsored by the pesticide registrant. The CARES model allows the user to construct a "canvas" (a graphical user interface with connected objects that represent functional modules and their relationship to each other) representing the sources of exposure (dietary - food, dietary - drinking water, and residential). Underlying data files, for example, carbaryl residue data in various foods, are specified within and utilized by the modules represented within the canvas. Figure 1 illustrates the canvas used for the dietary - food and drinking water modules or components of the canvas. Individuals within the CARES reference population (a statistically weighted sample of the U.S. Census; subpopulations from the U.S. Census can be selected from within the "Reference Population" module/icon in Figure 1) are assigned temporal (365-day) dietary (food and water) exposure profiles resulting from their consumption of foods (defined within the Water and Food Selector modules/icons in Figure 1) that may include those assumed to contain residues of carbaryl (defined within the Water and Food Match icons in Figure 1). For additional details, e.g., regarding the CARES Reference Population and other detailed aspects of the software's use and underlying methodology (e.g., development of 365-day temporal dietary profiles for individuals in the reference population), see the CARES Technical Manual available at http://cares.ilsi.org/CaresGuides.htm.

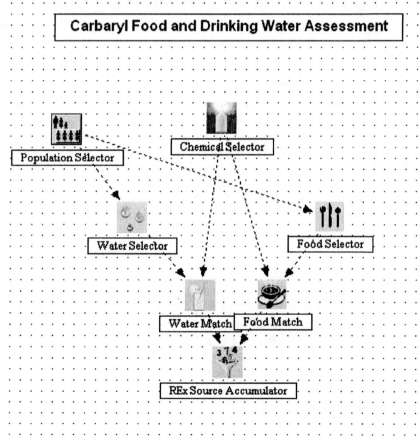

Figure 1. CARES canvas for the carbaryl aggregate human health risk assessment: dietary and water modules.

A CARES canvas was also developed for assessing potential residential exposures to adults who may apply carbaryl-containing consumer products, and those who may re-enter treated areas (e.g., lawns). The residential "exposure scenarios" in the assessment included lawn care (broadcast and spot treatment), vegetable garden care, ornamental care, tree care, and pet care. In the case of children, a CARES canvas was constructed to address plausible post-application scenarios, i.e., post-application re-entry onto broadcast-treated lawns, and post-application interaction with carbaryl-treated pets. These residential uses and associated potential exposure scenarios were based on a recent national survey of residential pesticide product use by U.S. households during a complete calendar year (12 month) period. The proprietary survey was conducted by the Residential Exposure Joint Venture (REJV), of which Bayer CropScience is a member company, and has been submitted and presented to the U.S. EPA's OPP Scientific Advisory Panel (SAP). The aggregate risk analysis presented in this chapter is based in part on the REJV national survey, and utilizes the complete 12-month data set (May 2001 – April 2002). This data set provides a fundamentally important foundation for the residential component of the carbaryl aggregate risk analysis (analogous to the use of national food consumption survey data as the basis for estimating potential dietary exposures to pesticide residues in foods). This dataset is an empirically-based, nationally representative profile of temporal (across a complete 12-month period) carbaryl product use in U.S. households. The profiles of use reported by participating households (HHs), include information regarding product-specific names and EPA Registration Numbers being applied, the dates of each application event, the method and site of each application event, and other ancillary information (e.g., which household member made the application and an indication of whether or not the product was disposed of following use). The HH-use profiles also provide actual co-occurrence of product applications, e.g., application of the same or different products containing a particular active ingredient, such as carbaryl, to one or more sites of application (e.g., lawns, ornamentals, fruit trees) on a given calendar day. Thus, the REJV survey provides a basis for identification of the specific carbaryl products used by U.S. consumers, their characteristics (e.g., the incidence of applying carbaryl liquid formulations to lawns by hose-end spraying equipment) and derivation of key conditional probabilities including day of week and month of year probabilities of use, and the likelihood of daily co-occurrences of two or more application events. It is important to note that the REJV survey provides statistical weights for each participating household to provide a means for relating the survey results to all U.S. households. The weighting scheme, which was designed in collaboration with EPA's OPP, is based on demographic criteria (proportionalities) from the U.S. Census that include geographic region, household income, household size, age of head of household, and household metropolitan statistical area size. However, un-weighted data from the survey can also be used to provide various proportional characterizations, e.g., the fraction of "carbaryl using households" that use a specific method of application such as a low-pressure sprayer, or the fraction of carbaryl using households applying carbaryl via a low pressure sprayer to specific sites such as lawns or vegetable gardens.

Daily residential exposures to individuals represented in the CARES reference population, are only estimated for those persons who have been assigned applicator or post-application exposure scenarios. These assignments are based on probabilities derived for carbaryl-specific product use from the REJV national consumer product use survey. The REJV-derived probabilities include the percentage (fraction) of households applying carbaryl-based products to specific sites of application (e.g., lawn). Additional conditional probabilities, for each site of application, are derived to determine assignment of the specific types of product being used (defined by the observed frequency for each method of application, for a given site of application). For example, based on the REJV survey results, lawn care broadcast applications can be made by using a granular formulation via a push spreader, or by using liquid formulations via either a hose-end sprayer or hand-pump sprayer. The probability associated with each of these methods determines the likelihood of a CARES individual using a specific method of application (adult applicator), given that they have been designated as being in a household using carbaryl on a given site (e.g., lawn). If the CARES individual is a child, only post-application exposures are estimated; whereas, if the CARES individual is an adult, both applicator and post-application exposures are estimated. Additional conditional probabilities derived from the REJV survey are used to define the likelihood of applying a product on any given day of the week, and month of the year, throughout the calendar year as represented in the CARES aggregate exposure simulation. Once a product-use event is assigned, applicator and post-application exposures on the day of application, and post-application exposures on subsequent days, are assumed to occur until residues dissipate or decline to negligible levels. Residential exposures, if assigned to a CARES individual are aggregated with that individual's estimated dietary (food and/or water) exposure, if relevant (i.e., if dietary exposure is estimated to occur on the same day as a residential exposure).

Daily dietary exposures are estimated for each CARES Reference Population individual based on their assigned food and water consumption diaries, for each day of a 365-day time period, and whether carbaryl residues are estimated to be present in the specific raw agricultural commodities and water sources represented in those daily diaries. The daily food consumption diaries or records used in CARES are from the U.S. Continuing Survey of Food Intake by Individuals (CSFII) 1994-1996 and 1998, and have been pre-assigned to each individual in the CARES Reference Population, for each day of the year, using a statistically-based matching procedure (see more detailed documentation at http://cares.ilsi.org/info/infolist.cfm?infoid=71. This procedure, and the overall CARES model, has been reviewed by the EPA's OPP FIFRA SAP.

As noted previously, carbaryl drinking water exposures have been incorporated into this aggregate risk analysis based on extensive drinking water monitoring data. However, it is important to emphasize that alkaline treatment conditions in municipal water treatment and the very rapid hydrolysis rates of carbaryl under alkaline conditions coupled with the low propensity to enter ground water make the potential for drinking water exposure very low. This is supported by an extensive monitoring database.

Professionally-applied residential products were not included in the assessment, e.g., treatment of residential ornamental or lawn sites by professional operators. In the case of professional operator applications, carbaryl represents a relatively small percentage as inferred from the California Department of Pesticide Regulation's (DPR) "California Pesticide Information Portal (CalPIP; http://jolie.cdpr.ca.gov/cfdocs/calpip/prod/main.cfm). The information in CalPIP's Pesticide Use Data section comes from the Use Report Transaction Record and various other data sets maintained by DPR. The Use Report Transaction Record contains information submitted by the grower or applicator about an instance of pesticide use. On an agricultural application, this includes what product was used, who used it, where the application was made, the commodity to which the application was made, when the application was made, and how much product was applied. When a Use Report Transaction Record is processed, information such as the chemical codes, chemical percent, and product information is retrieved. In the case of carbaryl's professional use for purposes of landscape maintenance and structural pest control (e.g., outdoor perimeter treatments), there were 228 application events reported in 2001 (across counties in California), out of a total of 65,537 total application events; thus, carbaryl professional applications represented only 0.35% of all applications for landscape maintenance and structural pest control. This suggests that carbaryl applications to residences represents a small percentage of total applications and that these associated potential post-application residential exposure events are not likely to contribute significantly to potential aggregate residential exposures. The remaining sections of this chapter summarize key considerations related to carbaryl's biological and physico-chemical characteristics, and a description of the methods and data used to estimate potential dietary (food and drinking water) and residential aggregate exposures.

Evaluation of Potential Aggregate Exposures to Carbaryl: Dietary and Non-Dietary Sources

Key non-occupational exposures considered in the carbaryl aggregate assessment included those resulting from dietary (food- and water-related contribution) and non-dietary (consumer products) sources. As noted previously, the CARES model provides a one year (12 months; 365 days) temporal profile of carbaryl exposure estimates for the entire U.S. population (which includes individuals who are exposed to carbaryl via diet or consumer product use, as well as those who are not). Thus, the aggregate exposure assessment represents a estimation of population-based or per capita temporal exposure profiles across individuals. Adults are defined as 18 years and older; and children ("toddlers" who may exhibit significant mouthing behavior and thus, experience potential incidental ingestion exposures) are defined as one to two years old. This toddler age range is typically selected to represent the subset of "children" who have higher consumption of certain foods and who may exhibit mouthing behavior (1) and other behaviors, e.g., crawling on environments such as residential lawns, that result in higher daily exposure potential.

The dietary, food-related assessment for carbaryl was based on residue data from crop field trials, the USDA's PDP, and a carbaryl-specific market basket survey (Carbaryl Market Basket Survey or CMBS). The water-related residues were based on a national water monitoring survey. The non-dietary sources are those resulting in potential exposures during and/or following the use of carbaryl consumer products on the outdoor residential application sites reported by households in the 12-month REJV national consumer pesticide product use survey. These included residential lawns (spot and broadcast treatment), vegetable gardens, ornamental plant care (flowers and shrubs), tree care, and pet care. Other registered uses of carbaryl were excluded because of negligible market share or use, e.g., mosquito control in residential areas.

Overview of the Dietary - Food Exposure Assessment Methods

The CARES dietary assessment presented here is based on refined food residue data and estimates of percent of crop treated from EPA's Biological and Economic Analysis Division (BEAD). The residue inputs were those recommended, in part, by EPA (2). The residue data were then imported into CARES using a pre-specified format (http://cares.ilsi.org/CaresTemplates.htm). The dietary assessment used the following residue data in this order of preference and availability.

1. Carbaryl Market Basket Survey (CMBS) data for the eight crops that were surveyed with surrogation to some other crop group members;
2. USDA Pesticide Data Program (PDP) data for crops for which it was available (through 1998) with some surrogation to closely related crop group members;
3. FDA monitoring data when CMBS or PDP data were not available; and

4. Field trial or tolerance data in the very few cases where appropriate monitoring data were not available.

Residue values were adjusted by the percent of crop treated according to U.S. EPA/OPP Standard Operating Procedure (SOP) No. 99.6 (Classification of Food Forms with Respect to Level of Blending). Percent of crop treated values were obtained from EPA's BEAD and represent data sources available through the year 1996. Residue values were also adjusted by processing factors from Bayer CropScience or literature studies where appropriate (2).

The CARES dietary and drinking water consumption data are based on the USDA FCID and the USDA CSFII 1994-1996 and 1998. The CSFII is a statistically representative survey of the food consumption of the U.S. population and subpopulations and has been conducted by the USDA since 1979. The CSFII records times and amounts of foods as eaten (e.g., pizza, apple pie). The FCID uses a set of "recipes" to translate these "foods as eaten" to raw and processed agricultural commodities (e.g., flour, tomatoes, and apples) for which residue data are available.

The CARES model uses a reference population of 100,000 people based on a Public Use Microdata Sample (PUMS) from the 1990 U.S. Census that is representative (when weighted) of the U.S. Population and subpopulations. The FCID individuals and their diets are matched to this reference population by a set of critical attributes. Using this matching procedure each individual in the reference population has a 365 day (1 year) consumption profile of raw and processed commodities summed for each day. Daily exposure is calculated for each individual in the selected population by moving through the consumption of each commodity for each individual for each day and randomly selecting a residue value from the appropriate residue distribution for that commodity, adjusting by any processing factors, and multiplying residue and consumption. The resulting commodity exposure values for that individual are summed for that day (an exposure-day). In the case of an "acute" or daily exposure assessment (24 hour period), the daily total values for each individual are placed in a population exposure distribution (distribution of exposure days). According to EPA acute dietary policy, the 99.9[th] percentile of this exposure distribution is selected for comparison to an acute Population Adjusted Dose (mg/kg of body weight/day; aPAD) which includes factors to address intra- and inter-species uncertainty.

Before residues were selected and calculations were made, a thorough evaluation of each redidue data source was performed. It is generally accepted that the best available residue data should be used in the dietary risk assessment, with the preference of use (from best to worst) in this order: Market Basket Monitroing > PDP Monitoring > FDA Monitoring > Field Trial Data > Tolerance. The more PDP and/or Market Basket monitoring data used in an assessment, the more realistic the resulting estimate of exposure. In particular FDA monitoring data may give a false picture of actual residues at the plate of the consumer. In most cases the FDA protocols do not include any home preparation or washing processes before analysis. The sample selection process may also be biased because of targeting trouble areas for sampling of the commodity.

USDA PDP monitoring data are obtained in distribution centers nationwide. The samples are subjected to typical home preparation procedures prior to analysis. The detailed PDP SOPs are available on their web site at www.usda.gov/science/pdp. The CMBS data were designed to get even closer to the consumer's plate, i.e., food as eaten by consumers. Foods were sampled randomly from grocery stores across the U.S. by shoppers mimicking typical consumers. Many discussions of typical home preparation, especially for fresh single-serving commodities which are to be eaten immediately, indicated, for example, that most people typically rub their fruits and vegetables as they wash them. Furthermore, if they are eaten directly after washing, they are then typically dried with paper towels or a cloth dish towel. The professional judgment of the study designers was to add this gentle rubbing, but not the drying process, to the protocol. Below is a direct comparison of the PDP and CMBS preparation protocols for one of the eight commodities tested.

Apple PDP
Wash for 15-20 seconds under cold tap water.
Air dry for at least 2 minutes on paper towels.
Remove stem.
Remove core with corer or cut in half or quarters and remove core.
Chop.

Apple CMBS
Wash for 15-20 seconds under cold tap water, rubbing gently.
Air dry for at least 2 minutes on paper towels.
Remove core.
Chop.

In most cases the protocols between PDP and CMBS are very similar. The only addition for the CMBS protocols is the gentle rubbing while washing. This is a reasonable assumption for the typical consumer, especially when the fresh fruit is to be immediately eaten or served. Bananas and oranges are, of course, peeled and most, if not all, carbamate residues would be removed with the peel. Many of the fresh fruits and vegetables used in the home may be further prepared by peeling, boiling, cooking, baking, etc. The fresh CMBS data are used for almost all food forms of a commodity including cooked forms (except where direct PDP data for the processed commodity exist). Extensive literature studies show that carbaryl residues are reduced from 25-100% with these types of home preparation procedures (Table II). Some of these home preparation and/or cooking factors are not accounted for in this assessment. Although the CMBS data account for some reasonable reduction in residue levels, the CMBS data probably represent a high-end level of residues which may be present in the typical American diet and are most appropriate for the aggregate assessment.

Decompositing of PDP, and especially FDA, monitoring data by the Allender method tends to give falsely high residue values at the upper tails of the imputed distributions (3). An examination of several decomposited carbaryl distributions shows some very high values, for example oranges (max. PDP value is 0.024ppm; max. decomposited residue is 1.68ppm) and pineapples (max. FDA value is 5.22ppm; max. decomposited residue is 35.78ppm.). These falsely high values, especially for the FDA decomposited distributions are likely to be driving exposure estimates at the upper ends of the distributions. Market Basket Studies and PDP single-serving monitoring studies have shown that there is essentially no difference in distributions of residues between single-serving and composite residues. Thus decomposition processes are not necessary. The significant effect of replacing single-serving distributions for decomposited PDP or FDA distributions is likely one of the reasons the CMBS data have a significant impact in lowering risk levels in the EPA series of analyses (2).

After considering all of the above information and the significance of the potential impact in reducing risk, a series of refinements were made for the dietary – food risk assessment in a step-wise fashion as explained below.

Step One involved the replacement of the residue data sources for leafy crops. Given that turnip tops, mustard greens, collards, and kale were all based on field trial data in the EPA analysis (2), PDP data were evaluated as an alternative. Specifically, SOP 99.3 supports surrogating the spinach PDP data to these crops after considering treatment patterns and percent crop treated. In support of surrogation with PDP data, an examination of "critical exposure contribution" from an earlier EPA dietary assessment using the CMBS data (1989-1992 CSFII) showed the following top contributors at the 99.9[th] percentile:

Infants:
- Canned Peaches, 37%
- Boiled Mustard Greens, 34%
- Pineapple Juice-Uncooked, 10%
- Boiled Turnip Tops, 9%

Children 1-6:
- Boiled Collards, 18%
- Boiled Turnip Tops, 17%

Table II. Home Preparation Process Factors for Carbaryl

Food	Process	Reduction	Reference
Broccoli	Cooking/washing	55%	8
Cabbage Heads	Cooking	90%	2
Cabbage Heads	Washing	75%	2
Cauliflower	Cooking/washing	94%	4
Grapes	Washing	49%; 85%	7
Green Beans	Canning	100%	11
Green Beans	Cooking/blanching	81%	11
Green Beans	Washing	52%	11
Okra	Cooking	42%; 25%	1, 14
Okra	Cooking/steaming	82%	1, 14
Okra	Washing	80%;66%;70%	1, 14
Onions	Washing	89%;98%;100%	9
Orchard Fruit	Washing	50%	12
Peas	Cooking/boiling	85%	3
Peas	Washing	70%	3
Spinach	Canning	99.5%	10
Spinach	Washing	70%	10
Tomatoes	Peeling/washing	99%	5, 6
Tomatoes	Puree/catsup	98%	5, 6
Tomatoes	Washing	66%; 68%, 84%	5, 6

1. Indian Journal of Plant Protection. 1996, 24, 86-89.
2. Pest Management and Econ. Zoology. 1994, 2, 131-134.
3. Plant Protection Bulletin. 1988, 40, 12-13.
4. Beitrage zur Trop. Land. Veter. 1982, 20, 89-95.
5. Indian Journal of Entomology. 1978, 40, 187-190.
6. Indian Journal of Entomology. 1973, 34, 31-34.
7. Indian Journal of Ag Sciences. 1978, 48, 179-183.
8. J. Ag. Food Chem. 1969, 15, 215-216.
9. J. Food Science Technology. 1978, 15, 215-216.
10. J. Ag. Food Chem. 1968, 16, 967-973.
11. J. Ag. Food Chem. 1968, 16, 962-966.
12. J. Assoc. Off. Anal. Che,. 1989, 72, 533-535.
13. Env. Health Criteria 1994, 153, 358pp.
14. Indian Journal of Ag Sciences. 1976, 45, 139-144.

- Strawberries, uncooked, 14%
- Pineapple Juice, uncooked, 6%
- Peaches cooked, 6%

Step Two involved the incorporation of 1999-2000 PDP data for newly sampled crops. A similar examination of the contributors for the analysis from Step One revealed that many of the top contributors were foods for which PDP data had recently become available (see Annual Summary for Pesticide Data Program 1999 and 2000 http://www.ams.usda.gov/science/pdp/Download.htm). No decomposition was incorporated for any of these commodities' food forms. The following data were incorporated in Step Two:

- Cucumber PDP (1999, 2000) replaced cucumber FDA data.
- Cherries PDP (2000) replaced cherries FDA data.
- Pineapple PDP (2000) replaced pineapple FDA (assumed 50% crop treated (imported) based on EPA's analysis).
- Strawberries PDP (1999-2000) replaced strawberry FDA.

Step Three involved the refinement of squash and peach data. A recent analysis of top contributors showed two additional contributors of significance that have more realistic residue data available. They are summer squash and canned peaches. The 1999-2000 cucumber PDP data were surrogated to summer squash (following SOP 99.3) which still used FDA data (no washing or peeling before analysis). The biggest contributor for infants was consistently canned peaches. It was found that the 1997 PDP data for canned peaches had been listed incorrectly in the 1997 PDP summary as fresh peaches (4). These data were substituted with the CMBS fresh peach data since they reflect the significant residue reduction when cooking/canning is done for peaches.

Based on the evaluation of each residue data source, a line-by-line examination of the residue source for each food and food form, and the step-wise refinements described above, the selection of commodities and associated residues were finalized. Table III below outlines new residue data sources for selected commodities listed alongside the EPA choices.

Overview of Dietary - Drinking Water Assessment Methods

Evaluation of Potential Water Residues: Use of Data from a National Surface Water Monitoring Survey for Carbaryl

The data from the carbaryl registrant (Bayer CropScience) monitoring study is the best source of information on exposure to carbaryl from drinking water

Table III. Data Sources for Selected Carbaryl Crops

Commodity	EPA Data Source	New Data Source
Beets, garden, roots	Beet Field Trial	Carrot PDP
Radishes	Beet Field Trial	Carrot PDP
Rutabagas	Turnip Field Trial	Carrot PDP
Turnip, roots	Turnip Field Trial	Carrot PDP
Endive	L. Lett. FDA	Spinach PDP
Swiss Chard	Celery PDP	Spinach PDP
Brussel Sprouts	Cabbage FDA	Lettuce MBS
Cabbage	Cabbage FDA	Lettuce MBS
Collards	Mustard Field Trial	Spinach PDP
Kale	Mustard Field Trial	Spinach PDP
Kohlrabi	Cabbage FDA	Broccoli MBS
Turnip, tops	Turnip Field Trial	Spinach PDP
Mustard Greens	Mustard Field Trial	Spinach PDP
Dried Beans and Peas	Dried Bean Field Trial	Soybean PDP
Succulent Peas	Peas FDA	Green Beans PDP
Eggplant	Sw. Pepper FDA	Tomato MBS
Peppers, chili	Hot Pepper FDA	Tomato MBS
All peppers	Pepper FDA	Tomato MBS
Cucumbers	Cucmuber FDA	Cucmber PDP
Summer Squash	S. Squash FDA	Cucumber PDP
Cherries	Cherries FDA	Cherries PDP
Rice	Rice Field Trial	Rice PDP
Pineapples	Pineapple FDA	Pineapple PDP
Strawberries	Strawberry FDA	Strawberry PDP

(5). This study consisted of weekly sampling of twenty community water systems located in areas with the highest use of carbaryl. Sixteen of these systems were located in areas of agricultural use while four were in suburban areas representing non-agricultural use. Agricultural locations were sampled weekly during the application season, and shortly thereafter switched to monthly sampling. The non-agricultural locations were sampled weekly throughout the study. Table IV provides the locations of the twenty community water systems and summarizes the analytical results.

In its evaluation, the EPA/OPP's Environmental Fate and Effects Division (EFED) has used PRZM/Exams simulations based on the "Index Reservoir." These simulations predict residues considerably higher than were measured in the registrant's monitoring study. The major reason for the higher model predictions using PRZM/EXAMS results are:

1. The assumed use of carbaryl in a watershed is two to three orders of magnitude higher than is actually occurring;
2. The Index Reservoir does not adequately represent the watershed of a community water supply system;
3. The residue levels predicted in the Index Reservoir do not adequately represent the residue levels measured at the inlet of a community water treatment facility, or in the potable water supply after treatment in a community water treatment facility;
4. The assumed use of carbaryl at the maximum application rate and for the maximum percentage of crop area within the watershed remains constant over the entire 36 year simulation period; and
5. The percent crop area values combined with product use at maximum application rates for 30+ consecutive years overestimate actual agricultural product use practices; and
6. Conservative parameter values also contribute to the over-prediction.

In a comparison between the values from the EPA modeling and the registrant study, although there is uncertainty associated with the peak values, the number of samples collected in the registrant study was sufficient to establish valid estimates of the 99th and lower percentiles of exposure in high carbaryl use areas.

The results of other recent drinking water monitoring programs (the USGS/EPA reservoir sampling program and USDA PDP program) showed carbaryl residues of lower magnitude with less frequency of detections than the registrant's drinking water study. This is probably because the registrant study was targeted toward carbaryl high-use areas. The USGS's National Water Quality Assessment (NAWQA) program (http://water.usgs.gov/nawqa/) which included sampling locations closer to treated fields had higher carbaryl residues and a greater frequency of detection. However, the maximum values observed in the NAWQA program were still two orders of magnitude lower than those in the PRZM/EXAMS modeling.

Implementation of Water Residue Monitoring Data In CARES

Generation of Daily Values

Since CARES requires daily values of exposure in calendar years, the data from the registrant drinking water monitoring study needed to be processed to give the necessary information. In the processing of these data the assumption was made that the concentration measured at a time point was constant until the next sampling point. This approach was used rather than a linear interpolation

Table IV. Summary of Results from the Carbaryl Drinking Water Monitoring Study.

| Site | Maximum Daily Concentration (ppt) | | | | | | Annual TWA Conc. in Outlet Water (ppt)[a] | | |
| | Inlet Water | | | Outlet Water | | | | | |
	1999	2000	2001	1999	2000	2001	1999	2000	2001
Manatee, FL	9	3	25	11	ND	19	1	1	3
West Sacramento, CA	3	24	14	3	10	9	1	1	1
Lodi, CA	12	31	4	4	7	ND	1	1	1
Riverside, CA	8	ND	ND	ND	NA	NA	1	1	1
Lake Elsinore, CA	ND	3	6	NA	NA	ND	1	1	1
Corona, CA	ND	ND	ND	NA	NA	NA	1	1	1
Beaumont, TX	ND	ND	ND	NA	NA	NA	1	1	1
Point Comfort, TX	18	5	ND	ND	ND	NA	1	1	1
Penn Yan, NY	ND	23	ND	NA	ND	NA	1	1	1
Westfield, NY	21	5	ND	ND	9	NA	1	1	1
Jefferson, OR	ND	10	4	NA	ND	ND	1	1	1
Coweta, OK	4	ND	ND	ND	NA	NA	1	1	1
Pasco, WA	2	3	ND	ND	ND	NA	1	1	1
Manson, WA	ND	ND	ND	NA	NA	NA	1	1	1
Deerfield, MI	10	4	22	160	ND	4	5	1	1
Brockton, MA	31	27	ND	ND	3	NA	1	1	1
East Point, GA	18	18	13	3	8	ND	1	1	1
Midlothian, TX	14	ND	14	ND	NA	ND	1	1	1
Cary, NC	4	ND	ND	ND	NA	NA	1	1	1
Birmingham, AL	23	35	40	ND	ND	32	1	1	2

[a] Annual Time Weighted Concentration, outlet values substituted for inlet values when available; values below the detection limit were considered to be half the detection limit (or 1 ppt).

ND Not detected.

NA No outlet samples analyzed due to carbaryl residues not being detected in inlet samples.

of the data with time, because it results in the higher concentrations being present for longer periods of time thus increasing the exposure at the higher percentiles (it also has the effect of slightly decreasing exposure at lower percentiles). The choice of a constant value or linear interpolation has no effect on estimates of chronic exposure. The decision to keep the concentration constant between the sampling points rather than from a period starting halfway between the sampling point and the previous sampling point through halfway between the sampling point and the following sampling point was based on simplicity. Since the sampling intervals were approximately constant during the period when residues were present, this simplifying assumption will have negligible impact.

The registrant studies were conducted for three years, but starting dates occurred in the spring of 1999 rather than on January 1 such that three complete years of data were collected during four different calendar years. A detailed description of data management and analysis methods used to create a daily time series of carbaryl residue levels in drinking water used in the CARES assessment are described in the next section.

In the generation of daily values, the values from finished water were used when they were analyzed (i.e. when raw water residues exceeded the limit of detection), otherwise the residue values for the raw water were used.

Building of the National Distribution

Because there was no apparent effect of geography on the results of the monitoring program, no stratification was performed in the input of the monitoring data into CARES. Instead, CARES randomly assigned one of the 60 annual time series (20 sites with three years per site) to each individual in the reference population. The effect of regional distributions was evaluated by including data from only one community water system in other CARES simulations.

The carbaryl water monitoring survey data file was imported into SAS™ (Release 8.02) as a comma-delimited text file. As noted previously, the original file contained the monitoring results for 20 sites in the U.S. that were selected as representative of watersheds with high carbaryl use. Samples were collected over a three-year period, generally covering a period from 1999 through 2002. Intervals between samples ranged from five to 49 days. Sampling at the study sites began February to June, 1999 and extended for approximately three years from the start date. Samples reporting non-detectable residue levels were assigned values of one-half the limit of detection (LOD), 1 ppt.

Each time series was first expanded to cover the four-year period from January 1, 1999 to December 31, 2002. Observations at the beginning of the time series (January to start date, 1999) that were not collected in the original time series were filled in with sample results from the same time period for the following year (January to start date, 2000). Similarly observations at the end of the time series (stop date to December 2002) that were not collected in the original time series were filled in with sample results from the same time period for the previous year (stop date to December 2001).

Once the complete four year period was created, missing observations between sample days were estimated using a step function (EXPAND procedure in SAS), where the observed residue value represented the first day of the step function, with the same residue value assigned to days with missing data until the next observed residue value. The expanded time series was then subset to cover exactly three one-year periods beginning at the start of sampling in 1999 and ending with a stop date exactly three years later. Therefore, each study site had 1096 observations, comprised of two periods of 365 days (1999 and 2001) and one period of 366 days (2000 leap year) which was truncated to 365 days. Each 1-year period began with a start date equal to the first sample results and continued for 365. The first year contained 366 days for start dates beginning after March 1, 1999, or 365 days for start dates beginning before March 1, 1999.

Since each one-year period covered two calendar years, the sample results were wrapped to create a single one-year period for each of three sample years. For example, a study site with a start date of February 17, 1999 resulted in three one-year periods: February 17, 1999 through February 16, 2000; February 17, 2000 through February 16, 2001; and February 17, 2001 through February 16, 2002.

The dates within each one-year period were assigned a Julian date to order the days from January 1 through December 31, regardless of year. Julian dates represent the ordered dates (month and day) within the year, assigning a value of "1" for January 1 and "365" for December 31 with all dates in between also numbered consecutively, for years that are not leap years. Each study site is described by a time series containing 1,096 observations, two years at 365 days and one year at 366 days. The leap year (2000) was truncated to 365 days to standardize the time series and simplify the analysis by deleting the carbaryl drinking water residue value for December 31, since water samples were collected at some study sites on February 29, but were not collected on December 31. Thus, a total of sixty, 365-day data sets were available for use in the CARES drinking water exposure assessment simulation. Each individual from the CARES reference population included in the simulation was assigned a complete 365-day time series of carbaryl water residues based upon random selection from the pool of sixty possible data sets.

Daily drinking (tap and non-carbonated bottled) water consumption used in the CARES drinking water exposure assessment was obtained from the CSFII/FCID data comprising the individual's daily consumption profile.

Overview of Non-Dietary Residential Exposure Assessment Methods

Potential non-dietary, residential exposure scenarios were estimated for the following consumer products and scenarios based on temporal product use patterns reported in the 12-month REJV national consumer product pesticide use survey (EPA MRID# 46099001):

1) Lawn (turf) Care (CARES Scenarios 101 – broadcast, and 117 – spot treatment)
   a. Dust (2 to 10% AI; pour/shake; spot treatment)
   b. Ready-To-Use (RTU) (includes spritz, trigger, aerosol and direct pour methods; spot treatment)
   c. Concentrates, Hose-End Sprayer (broadcast treatment)
   d. Concentrates, Handwand/Pump Sprayer (broadcast treatment)
   e. Granular, Pellets (push-spreader; broadcast treatment)
2) Vegetable Garden Care (CARES Scenario 102):
   a. Dust (2 to 10% AI; pour/shake);
   b. RTU (includes spritz, trigger, aerosol and direct pour methods; spot treatment)
   c. Concentrates, Hose-End Sprayer (broadcast treatment)
   d. Concentrates, Handwand/Pump Sprayer (broadcast treatment)
3) Ornamental Plant Care (CARES Scenario 103)
   a. Dust (2 to 10% AI; pour/shake);
   b. RTU (includes spritz, trigger, aerosol and direct pour methods)
   c. Concentrates, Hose-End Sprayer
   d. Concentrates, Handwand/Pump Sprayer
4) Tree Care (CARES Scenario 104)
   a. Dust (2 to 10% AI; pour/shake);
   b. RTU (includes spritz, trigger, aerosol and direct pour methods)
   c. Concentrates, Hose-End Sprayer
   d. Concentrates, Handwand/Pump Sprayer
5) Pet Care (CARES Scenario 109)
   a. Dust (5 to 12.5% AI; pour/shake)

The exposure assessment methods selected in CARES for each scenario noted above are further characterized in Table V below with respect to whether applicator and/or post-application pathways are considered, the relevant exposure route(s), subpopulations involved, and specific exposure assessment methods including the CARES method/algorithm identification number (see further details in the CARES Technical Manual available at http://cares.ilsi.org/CaresGuides.htm).

Table V. Residential Scenarios and Assessment Methods Used in the CARES Aggregate Assessment for Carbaryl

CARES Scenario and Number	Pathway	Route	Receptor[a]	Assessment Method (CARES algorithm number)
Lawn Care (broadcast); 101	During Application	Dermal	Adult	Unit Exposure, Area Treated (Dermal 101)
		Inhalation	Adult	Unit Exposure, Area Treated (Inhalation 101)
	Post Application	Dermal	Adult, Child	Transfer Coefficient, Residue (Dermal 103)
		Ingestion	Child	Mass Balance (Ingestion 107)
Lawn Care (spot treatment); 117	During Application	Dermal	Adult	Unit Exposure, Area Treated (Dermal 101)
		Inhalation	Adult	Unit Exposure, Area Treated (Inhalation 101)
Vegetable Garden Care; 102	During Application	Dermal	Adult	Unit Exposure, Area Treated (Dermal 101)
		Inhalation	Adult	Unit Exposure, Area Treated (Inhalation 101)
	Post Application	Dermal	Adult	Transfer Coefficient, Residue (Dermal 103)
Ornamental Plant Care; 103	During Application	Dermal	Adult	Unit Exposure, Area Treated (Dermal 101)
		Inhalation	Adult	Unit Exposure, Area Treated (Inhalation 101)
Tree Care; 104	During Application	Dermal	Adult	Unit Exposure, Volume Applied (Dermal 102)
		Inhalation	Adult	Unit Exposure, Volume Applied (Dermal 102)
Pet Care; 109	During Application	Dermal	Adult	Unit Exposure, Volume Applied (Dermal 102)
		Inhalation	Adult	Unit Exposure, Volume Applied (Inhalation 102)
	Post Application	Dermal	Adult, Child	Transfer Coefficient, Residue (Dermal 103)
		Ingestion	Child	Mass Balance (Ingestion 107)

[a] Adults are defined as individuals 18 years or older. Children are defined as ages 1 to 2.

Scenario-specific input variable values associated with the CARES residential assessment methods are not provided as part of this chapter, but were similar to those used in EPA's preliminary carbamate cumulative risk assessment, which included carbaryl. Details regarding residential input variable values can be obtained from documents developed for an EPA SAP meeting by following the links at http://www.epa.gov/pesticides/cumulative/carbamate_cumulative_factsheet.htm Additional carbaryl-specific residential input variable values are discussed in the EPA re-registration documents at http://www.epa.gov/pesticides /reregistration/carbaryl/. Key input data related to the residential "Event Allocation" module of CARES are discussed below. These data provide the basis for determination of product application and exposure event probabilities across the calendar year for each individual from the CARES Reference Population included in the simulation. Adults are defined as individuals 18 years or older. Children are defined as two different cohorts, ages one to two, and three to five years. Three to five year-old toddlers are expected to be more likely to engage in post-application, re-entry activities; however, both age groups may experience incidental dietary exposure and incidental ingestion exposure via "hand-to-mouth" behavior; thus, both age groups were addressed in the context of evaluating potential aggregate exposures.

Residential Scenario/AI Probabilities and Product-Specific Market Share Values

In the context of the CARES residential module, the "Scenario / AI Probability" represents the fraction of U.S. households (and individuals within those households) that are estimated to actually use or apply carbaryl-based consumer (or professional) products or who engage in post-application exposure-related activities (such as re-entry onto treated residential turf). In determining the Scenario / AI Probability, users of carbaryl products for any scenario (e.g., lawn care, vegetable garden care, etc.) are also differentiated from non-users. In the REJV 12-month survey, 101,566 U.S. households (HH) were initially asked to respond to a "screener questionnaire" that included questions regarding pesticide use status (use of any pesticide product in the past 12 months or plan to use pesticides in the next 12 months) and a variety of demographic information. Of this group of U.S. households, 70,427 provided questionnaire responses. Of this subset, 47,274 households classified themselves as pesticide users, i.e., 67 percent of the households surveyed, indicated that they had used pesticides in the past 12 months, or plan to use pesticides sometime during the next 12 months. Of the 70,427 households who responded to the demographic screener, 15,991 agreed to provide 12-month diaries of their pesticide use.

Some of these households completed all 12 months of the survey, while others completed a fraction of the year. All households that provided complete 12 month product use diaries were assigned statistical weights to relate them to the overall U.S. population based on demographic characteristics.

Based on records in the REJV 12-month survey's "Application" file (a file containing application information for all households who applied any pesticide product during their participation in the survey), it was determined, for example, that approximately 1.23% of U.S. households (based on un-weighted survey data) made carbaryl-based product applications to their lawns during a one-year period. The carbaryl lawn care products that comprise the 1.23% Scenario/AI Probability estimate, include those used for spot and broadcast treatment. Spot treatments (hand-held RTU products) are expected to result in negligible post-application exposure potential. Therefore, it is necessary to differentiate them from the proportion of applications that could be broadcast treatments (e.g., hose-end sprayer, handwand/pump sprayer, rotary spreader). This can be based on the proportions of RTU versus broadcast application methods.

The REJV 12-month survey indicates that 66% of the 1.23% carbaryl lawn care scenario probability (i.e., 66% of carbaryl lawn applications) are spot-treatment-related, whereas 34% are broadcast treatment-related. Therefore, the spot treatment lawn care products are assigned a Scenario/AI Probability (fraction) of 0.0123 x 0.66 = 0.0081; the "custom" scenario available in the CARES residential module was used for estimating potential carbaryl lawn spot-treatment-related exposures (adult consumer applicator exposure estimation only). The CARES residential module's "lawn scenario – 101" was used for carbaryl lawn broadcast products (adult consumer applicator, and adult and child post-application exposure estimation) based on a Scenario/AI Probability of 0.0123 x 0.34 = 0.0042.

It is important to note that in addition to the REJV survey, the EPA's National Home and Garden Pesticide Use Survey (6) provides household pesticide inventory and product use data (by target pest, application site, and application method); however, it does not reflect the current consumer product application practices for carbaryl. Further, NHGPUS does not provide any information regarding the co-occurrence of product use (e.g., the use of a carbaryl product on two or more sites on a given day).

Subsequent to determining the Scenario/AI Probabilities, each product type within a scenario category must be assigned a "market share." Thus, within a scenario such as lawn care, the respective market share of each product type that could be selected for use is assigned a fraction, and the sum of these fractions (across all products within a scenario) will equal "1." The "market share" is based on the incidence of each product type/application method within the REJV Application file, subset for carbaryl and a specific scenario (site of application). In the case of carbaryl lawn care for example, five types of products (differentiated by method of application) were reported as being applied: dusts, Ready-To-Use (RTU) sprays, liquid concentrations via hose-end sprayers, liquid concentrates via hand-pump sprayers, and granular/pellet application via push-spreaders. Thus, 1.23% of the U.S. population is assumed to use one of these products, one or more times per year (66% being spot treatment-related, and 34% broadcast treatment-related). As noted above, within scenario product market shares were derived from the REJV survey, based on the reported frequency of application of each product type within a given site of application or scenario (e.g., lawn).

Month-of-Year and Day-Of-Week Application Probabilities

CARES calendar-based modeling requires estimates of the likelihood that carbaryl will be applied during any given month during a one year period (to address potential seasonality of applications), and on any given day of the week (to address potential differences in the likelihood of applying carbaryl products for a given scenario on weekdays versus weekends). These probabilities can be readily derived from the REJV survey Application data file. This is accomplished by subsetting the Application file by a relevant site of application (e.g., Lawn? = yes) and carbaryl's PC Code, i.e., 50185. A "unique values" function is then performed on the "Month Applied" and "Day Applied" columns to determine frequency (and associated fraction or proportion) of each month (January through December) and each day of week (Sunday through Saturday), respectively.

Frequency of Carbaryl Product Applications Per Year

To determine the average number of applications of carbaryl products per year, within a given scenario, the following procedure was used. First, the REJV 12 month Application file was subset to a relevant site of application and carbaryl's PC Code (e.g., Lawn = yes and pc_code_1 through 6 = '056801'). Then, this file was subset by specific method(s) of application. Next a "unique values" function on HH ID was performed , and the incidence (frequency) that each unique value occurs was sorted and tallied. This action provided the frequency of applications per year, within a given scenario, by each method of application. For example, typically only one or two carbaryl lawn-related applications were made per year as reported in the REJV national survey.

Scenario Co-Occurrence Probabilities

In addition to Scenario/AI Probabilities, it is necessary to determine the likelihood that two or more residential exposure scenarios (and associated product use) may co-occur during a toxicologically relevant time period, i.e., in the case of carbaryl, one day (24 hour period). The REJV survey was designed to provide directly relevant data for this purpose. The procedure for deriving the scenario co-occurrence matrix is as follows. First subset REJV 12 month Application file for each relevant site, (e.g., Site (Lawn?) = yes and PC CODE = '056801' (carbaryl)). Then, based on the applications reported for one of the relevant sites (e.g., 145 lawn applications were reported), the incidence union ("U") was determined for lawn and vegetable garden, lawn and ornamentals (flowers, shrubs, and other ornamentals), lawn and fruit and nut trees, and lawn and pets. Finally, the respective incidence values are then represented as fractions of total applications to the lawn as the index site. This process is repeated for each of the other sites when they are represented as the index site. The resulting co-occurrence probability (fraction) matrix is derived as shown in Table VI. Thus, for example, the carbaryl co-occurrence probability matrix indicates that if a lawn care application is made at a residence, there is a 31% probability that an ornamental plant care application will also be made on the same day.

Table VI. CARBARYL: Scenario Co-Occurrence Probability Matrix

Exposure Scenario	LC	VGC	OPC	TC	PC
LC = Lawn Care (spot or broadcast)	0.000	0.117	0.310	0.034	0.034
VGC = Vegetable Garden Care	0.026	0.000	0.230	0.068	0.011
OPC = Ornamental Plant Care[a]	0.072	0.248	0.000	0.076	0.008
TC = Tree Care	0.034	0.306	0.320	0.000	0.000
PC = Pet Care[b]	0.100	0.120	0.080	0.000	0.000

[a] OPC includes ornamental flowers, shrubs/bushes, and other ornamentals; i.e., to estimate fractions, within the Application file, the union ("U")/incidence of lawn and flowers, and/or shrubs/bushes, and/or other ornamentals was determined.

[b] PC includes dogs, cats, and other pets

Carbaryl Dose-Response Evaluation

Determination of the most relevant "exposure metrics" (e.g., acute daily total absorbed doses, subchronic seasonal average daily absorbed doses, lifetime average daily absorbed doses) for comparison to dose-response relationships is informed by what is known about a given chemical's toxicokinetics and toxicodynamics. Carbaryl is an N-methyl carbamate that functions as a reversible acetylcholinesterase inhibitor (7). Carbamates, such as carbaryl, have been widely studied in this regard (7). For example, a large group of cholinesterase-inhibiting carbamates have undergone various phases of clinical trials in humans as treatments for Alzheimer's dementia (8) such as gangstigmine (9), rivastigmine (10), heptylphysostigmine (11) and eptastigmine (12). Physostigmine, the oldest cholinesterase-inhibiting carbamate that continues to be used in treatment of glaucoma was the prototype compound tested for Alzheimer's (13). All of these pharmaceutical compounds were chosen because they have a much longer half life, and thus time of effect, in humans than carbaryl. Coincidentally, all of these compounds are also more potent inhibitors of cholinesterase than carbaryl.

It is well known that carbamates are quickly metabolized and that cholinesterase inhibition is rapidly reversible, unlike the organophosphates (7). There are numerous examples of a divided dose of carbamates producing less cholinesterase inhibition than the sum of the dosage administered as a bolus. One such example comes from the literature on pre-clinical trials of the Alzheimer treatment candidate heptylphysostigmine (11). In addition to pharmaceuticals, there is a wealth of information about carbamate insecticides that have been tested in animals and humans for pharmacokinetics and pharmacodynamics (14, 15, 16, 17, 18, 19). Pesticides with very short half-lives (e.g., carbamates) that have reversible toxicological effects should be tested in a manner commensurate with exposure duration and frequency.

In addition to the rapid metabolism (the predicted metabolic pathway of carbaryl in humans is illustrated in Figure 2), the inhibition of cholinesterase produced by carbaryl is readily reversible. In humans and rats the half-life for cholinesterase inhibition *in vivo* is 2.6 and 3.0 hours, respectively (20). Metabolic profiles in rats and humans are similar in that they produce qualitatively the same major metabolites, i.e., hydrolysis of the carbamate linkage predominates and ring hydroxylation of intact carbamate occurs in both species to a lesser degree. Thus, rat toxicological endpoints and site of action are valid surrogates for humans. The metabolism, pharmacokinetics and pharmacodynamics of carbaryl require unique consideration for risk assessment, and as a result, the traditional method of estimating a Margin of Exposure (MOE; expressed as total daily dosage at the No-Observed-Adverse-Effect-Level or NOAEL divided by an estimated total daily exposure for a given human cohort) is not as relevant. Sufficient inhibition of brain cholinesterase can result in adverse effects (7). Thus, the magnitude of short-term cholinesterase inhibition can be most closely associated with peak or plateau tissue levels in target tissues such as brain (21, 22). This suggests the need for an alternative to the classic MOE calculation such as estimating the ratio of the peak brain levels at the oral, systemic absorbed dose NOAEL (i.e., 1 mg/kg/day) to the peak (following one exposure event) or plateau (following repeat exposure events) brain levels estimated for relevant human absorbed dose levels from biological monitoring studies of specific human cohorts which would reflect aggregate dose levels. Thus, physiologically-based pharmacokinetic and dynamic modeling represents a more relevant and refined approach in the case of a chemical such as carbaryl, that the more common daily exposure metric. Additional research and model development in this regard is being pursued specifically for carbamates such as carbaryl, including the development of a less than 24-hour time step in the CARES model to allow for accommodation of the reversibility kinetics of cabamate-related cholinesterase inhibition associated with intermittent exposure events (e.g., food eating occasions) that may occur within a day (24 hour period).

EPA has evaluated the toxicity profile of carbaryl and has recommended toxicity endpoints and associated exposure metrics for risk assessment purposes (*23, 24*). The short-term exposure metric (one to seven days of exposure) and intermediate-term exposure metric (seven days to several months) for carbaryl human health risk analyses are both based on the cholinesterase inhibition effects of carbaryl. The short-term, systemic oral No-Observed-Adverse-Effect-Level (NOAEL) of 1 mg/kg/day is based on maternal toxicity observed in an oral developmental neurotoxicity study conducted in rats. The Lowest-Observed-Adverse-Effect-Level (LOAEL) in the study was 10 mg/kg/day based on RBC, plasma, and brain cholinesterase inhibition. In addition, there were decreased body weight gains and alterations in Functional Observational Battery (FOB) measurements at the LOAEL. The intermediate-term oral NOAEL was also 1 mg/kg/day based on an oral subchronic neurotoxicity study in rats. The LOAEL was 10 mg/kg/day based on RBC, plasma, and brain cholinesterase inhibition and changes in the FOB measurements. It is important to note that the NOAELs for the short-term and intermediate-term assessments are the same and are based on similar observations from the developmental neurotoxicity and subchronic neurotoxicity studies. The short-term oral NOAEL of 1 mg/kg/day was also recommended by EPA for the inhalation route (*24*).

In the case of the dermal route, a systemic applied dose NOAEL of 20 mg/kg/day was identified from a repeat-dose (4-week) dermal toxicity study in rats with technical carbaryl. The LOAEL in this study (i.e., 50 mg/kg/day) was based on decreased RBC cholinesterase in males and females and brain cholinesterase in males.

Thus, route-specific NOAELs (oral = 1 mg/kg/day; inhalation = 1 mg/kg/day; and dermal = 20 mg/kg/day) are compared to route-specific exposure estimates to derive route-specific Margins of Exposure (MOEs). The

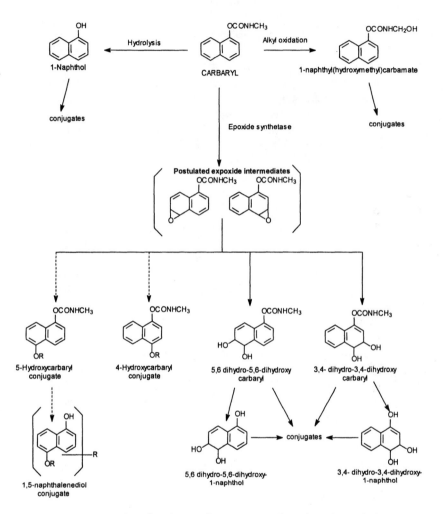

Figure 2. Carbaryl's chemical structure and predicted metabolic pathway in humans.

total or aggregate MOE can then be estimated as *[1 / ((1 / oral MOE) + (1/ inhalation MOE) + (1/dermal MOE))]*. The CARES model actually derives the total MOE in a manner that is equivalent to the above equation, based on a relative potency factor (RPF) or Toxic Equivalent Factor (TEF) approach as described by EPA in their organophosphate cumulative risk assessment (see U.S. EPA/OPP's "Preliminary OP Cumulative Risk Assessment" (http://www.epa.gov/pesticides/cumulative/pra-op/) and related documentation (http://www.epa.gov/pesticides/cumulative/pra-op/hedinput.htm). Within the CARES model's "Toxicology Parameters" file, the route-specific NOAELs (or benchmark doses - BMDs) are listed, with the respective NOAELs expressed in units of mg/kg/day. The user then selects the route-specific NOAEL to be used as the "index" NOAEL or benchmark, for example, the dermal short-term NOAEL of 20 mg/kg/day. When deriving the total aggregate MOE, the CARES model will then express all exposure estimates as "dermal route equivalents" or "Toxic Equivalent Doses" (TEDs) based on RPF adjustments for the dermal exposures versus oral and inhalation route exposures when the dermal route is selected as the index NOAEL. The total daily TED or total daily dermal equivalent exposure for a given person (*i*) on a given day (*j*) in the CARES Reference Population is then defined as: *(daily dermal exposure$_{i,j}$ x RPF$_{dermal}$) + (daily oral exposure$_{i,j}$ x RPF$_{oral}$) + (daily inhalation exposure$_{i,j}$ x RPF$_{inhalation}$).*

The daily aggregate MOE (for each person-day, for each individual in the user-specified CARES Reference Population, or sub-population) is then estimated using the index NOAEL (e.g., 20 mg/kg/day divided by the TED as total dermal equivalent exposure, across all routes).

In the case of carbaryl, if the dermal short-term NOAEL of 20 mg/kg/day is selected as the index NOAEL for RPF derivation and subsequent route-specific exposure adjustments, then the RPF$_{dermal}$ = 1, and the RPF$_{oral}$ and RPF$_{inhalation}$ = 20 (given that the oral and inhalation short-term NOAELs are 1 mg/kg/day). The aggregate MOE would then be estimated as 20 mg/kg/day / dermal-based TED (across all routes).

Because carbaryl is a reversible cholinesterase inhibitor (7), there is no observed cumulative effect on cholinesterase enzyme activity inhibition with increasing duration of carbaryl oral administration. This pattern is expected for the N-methyl carbamates, as opposed to the organophosphates, and indicates that the enzyme inhibition has completely reversed prior to the next day's oral administration of carbaryl. Thus, it can be concluded that a chronic, non-cancer risk assessment is not required because the carbaryl non-occupational exposures are not likely to produce long-term exposures of several months to a lifetime. The oral NOAEL from the chronic dog toxicity study further illustrates the rapid reversibility of cholinesterase inhibition by carbaryl. The NOAEL for plasma cholinesterase inhibition was 1.4 mg/kg/day. This chronic oral NOAEL is essentially the same as the short-term and subchronic oral NOAELs for cholinesterase inhibition, and it provides further evidence that the cholinesterase inhibition potential of carbaryl is reversible and not significantly influenced by the frequency of exposure.

Dermal absorption data are available for carbaryl. A dermal absorption study in rats with the carbaryl XLR 43.9% AI formulation was evaluated by EPA's Office of Pesticide Programs and used to estimate the dermal absorption potential of carbaryl. A dermal absorption adjustment factor of 12.7% was selected by the Agency based on the percentage of the lowest dose administered that was absorbed after 10 hours of exposure. Thus, an estimate of the absorbed dermal dose can be made by adjusting dermal exposure estimates from CARES output files (adjustments made to output files/results by the model user after running the CARES model) by the 12.7% dermal absorption factor. The use of the .12.7% dermal absorption factor will likely overestimate the dermal absorbed dose of solid formulations of carbaryl because the dermal absorption factor was obtained from a liquid formulation. The dermal absorption potentials from solid formulations are typically 10-fold or more lower than the dermal absorption potential from a liquid formulation. For the purpose of this assessment, the dermal absorption potential for the granular and dust formulations is assumed to be equal to the liquid formulation, or 12.7%. Using 12.7% is conservative, because for most chemicals, rats typically absorb an average of five times as much as humans through the skin (25). This estimate of absorbed dermal dose can then be used to compare predicted dose estimates to measured doses observed in biomonitoring studies.

Similarly, oral and inhalation absorbed doses can be estimated based on adjustment of the oral and inhalation exposures predicted by CARES. CARES exposure output from this assessment can be conservatively adjusted to "absorbed dose" estimates, by assuming that 100% of carbaryl oral and inhalation exposures are systemically absorbed. Thus, to compare CARES distributional output for route-specific exposure to available biological monitoring data for carbaryl, dermal, oral and inhalation exposures can be adjusted to estimate total absorbed daily dose using the following factors: 0.127 for dermal, and 1 for oral and inhalation. As noted above, the total MOEs derived in this assessment were based on route-specific exposure and route-specific NOAELs, in contrast to derivation of a total MOE based on total absorbed dose estimates and a systemic, absorbed dose NOAEL.

It is important to note that because carbaryl is a rapidly reversible cholinesterase inhibitor, and has a short plasma half-life (20), potential aggregate health risks may be overstated by not quantitatively addressing the chemical's toxicokinetics and toxicodynamics. For example, uptake or absorption rate kinetics for the oral and dermal routes into the blood/plasma compartment of the body are significantly different. The oral administration of carbaryl leads to a more rapid influx of carbaryl into blood/plasma compared to the slower absorption that occurs from the dermal route of exposure. The slower absorption of carbaryl through the skin, as occurs with residential or occupational exposure permits the body to effectively de-activate the carbaryl molecules (i.e. through reversibility of cholinesterase binding or metabolism) as they are absorbed over time.

Finally, while the Agency has determined that carbaryl is a possible carcinogen and has estimated a Q* of 1.19×10^{-2} (mg/kg/day)$^{-1}$ for carbaryl (26), the assessment presented herein focuses on non-cancer endpoints, given that they are more relevant to the short- and intermediate-term aggregate exposure patterns likely experienced by consumers in the residential setting.

Aggregate (Per Capita) Exposure Assessment and Risk Characterization

Background

The relative safety of a given exposure is often reported as the Margin of Exposure (MOE); the MOE is defined as the route-specific NOAEL divided by route-specific exposure (or the systemic NOAEL divided by the total absorbed dose). Overall aggregate MOEs can be derived from route-specific MOEs using the following equation: [1 / ((1 / oral MOE) + (1/ inhalation MOE) + (1/dermal MOE))]. This equation can also be represented by the use of a relative potency factor (RPF) approach as described by EPA in their organophosphate cumulative risk assessment (27, 28). Further, the RPF or Toxic Equivalent Factor (TEF) approach can be used for aggregate (single chemical, multiple routes) or cumulative (multiple chemicals, multiple routes) exposure and risk estimation.

In most cases, an MOE of 100 or greater is associated with an acceptable level of safety; this reflects the magnitude of the assigned Uncertainty Factor (UF). The UF results from multiplication of a series of numerical factors (usually 10 each) that rate the relevance of the time-frame of the study relative to the time-frame of the exposure in which one is interested; the uncertainties of extrapolation from laboratory animal data to human exposures; and inter-individual variation in humans.

Potential Aggregate Health Risks Associated with Carbaryl

Table VII presents the aggregate acute (single day) MOEs for carbaryl associated with potential daily exposures resulting from dietary (food-related

Table VII. CARES Carbaryl Aggregate Risk Assessment Summary Results: Dietary (food-related), Drinking Water, and Residential (consumer-related)

Aggregate – Children, 1-2 years

Receptors	Children 1 to 2 years		
Subpopulation	3,367 in CARES Reference Population		
Number of Exposure Days[a]	1,228,955 (= 3,367 x 365)		
Percentile (%)		*99.9th*	*99th*
Rank Order in Sorted Output		1,230	12,291
Daily Exposure (oral equivalents; mg/kg/day)		0.008	0.0035
MOE		125	288

Aggregate – Adults, 20-49 years

Receptors	Adults 20 to 49 years		
Subpopulation	33,538 in CARES Reference Population		
Number of Exposure Days	12,241,370 (= 33,538 x 365)		
Percentile (%)		*99.9th*	*99th*
Rank Order in Sorted Output		12,243	122,415
Daily Exposure (oral equivalents; mg/kg/day)		0.0027	0.0008
MOE		363	1,242

[a] Represents the total number days included in the simulation across all persons in the selected cohort.

and drinking water) and non-dietary (residential consumer products) sources. Additional detailed results are presented in Tables VIII-XII for adults and children (1-2 years). Results for children 3-5 years old, while not shown, are similar to the 1-2 year olds. MOEs are population-based or per capita percentiles (in contrast to "product users only"). Thus, the MOE percentiles presented are for the U.S. population and include non-exposed individuals, as well as those individuals in households that are exposed to carbaryl through their diet or consumer product use. Thus, the MOEs can be characterized as "Per Capita" or population-based estimates that include exposed and non-exposed individuals.

The 99.9th and 99th percentile aggregate MOEs for children aged 1-2 years are 125 and 288, respectively. The 99.9th and 99th percentile aggregate MOEs for adults aged 20 to 49 years are 363 and 1242, respectively. Based on Tables VIII-X, dietary (food) exposures are the primary contributor to the aggregate MOE for both age groups.

Conclusions and Uncertainties

The aggregate and route-specific risk estimates (MOEs) presented in Table VII indicate that the estimated acute MOE distributions expressed on a per capita basis are acceptable, i.e., greater than 100. The per capita MOE values at the 99.9th percentiles are presented separately in Tables VIII-X for dietary – food, drinking water, and residential sources. The overall aggregate MOEs (dietary – food + drinking water + residential) are predominantly influenced by the dietary – food exposures; thus, the dietary – food MOEs presented also represent the approximate overall aggregate MOEs, since drinking water and residential exposures do not significantly contribute to the overall MOE.

The dietary – food MOE values at the 99.9th percentiles (using refined RED assessment inputs) were estimated to be approximately 365 and 126 for adults (20-49 years) and children (1-2 years), respectively. The residential MOE values at the 99.9th percentile were estimated to be approximately 4,300 and 8,400 for adults (20-49 years) and children (1-2 years), respectively; and those for drinking water were estimated to be >100,000 for adults and children. The residential MOEs at the 99.9th percentile reflect the relatively small proportion of U.S. households impacted by carbaryl residential product use in the context of a population-based, aggregate human health assessment. Given the conservative biases inherent in aspects of this assessment, the MOEs at higher percentiles (e.g., at and above the 99.9th percentile) are likely to represent exposure profiles that are very rare situations.

In the case of the dietary – food component of the carbaryl aggregate assessment, incorporating the refinements outlined above in the CARES inputs resulted in an estimated 99.9th percentile exposure for children, age 1-2 years, of 0.0078 mg/kg bw/day or 78% of the acute RfD. The Adults 20-49 subpopulation exposure is 0.00274 mg/kg bw/day, or 27% of the acute RfD. Thus, a significant reduction in risk can be obtained when incorporating residue data that are more relevant, of higher quality, and are more contemporary.

Examination of the top contributors from the refined dietary analysis (see Tables VIII and IX) reveals that the major contributors to risk are now strawberries, blueberries, and cooked peaches. The source of the EPA blueberry data is FDA monitoring data with some very high residues. All of the caneberry residues used in the assessment are FDA data. Blueberries have been discussed as a commodity to be included in future PDP surveys. The PDP data would almost certainly show much lower residue levels (if any) in blueberries

Table VIII. Detailed Results Dietary – Children, 1-2 Years

Children			1 to 2 years old	
Number in Subpopulation			3,367	
Number of Exposure Days			1,228,955	

CARES Refined Assessment (Residues: Actual Values)				
Percentile (%)	Exposure (oral equivalent; mg/kg/day) (Wighted per Capita)	MOE (Weighted per Capita)	Foods Resulting in Exposure >0.01 mg/kg/day	Number of Occurences
			Strawberry	238
99.9	0.0078	126	Blueberry	69
99.8	0.0061	158	Peach	41
99.7	0.0051	191	Strawberry, juice	36
99.6	0.0046	211	Apple, juice	26
99.5	0.0043	227	Grape	26
99.4	0.0040	243	Nectarine	22
99.3	0.0038	256	Olive	21
99.2	0.0037	267	Blackberry, juice - babyfood	19
99.1	0.0036	278	Raspberry, juice	19
99.0	0.0035	288	Bean, snap, succulent	15
98.9	0.0033	298	Raspberry	14
98.8	0.0032	308	Beef, meat byproducts	14
98.7	0.0031	317	Peach, juice	9
98.6	0.0031	326	Blueberry- babyfood	9
98.5	0.0030	334	Beef, kidney	9
98.4	0.0029	341	Grape, raisin	7
98.3	0.0029	348	Cherry- babyfood	6
98.2	0.0028	354	Peach, dried	5
98.1	0.0028	361	Cherry, juice	4
98.0	0.0027	367	Bean, cowpea, succulent	2
97.9	0.0027	374	Apricot, dried	2
97.8	0.0026	380	Cherry	2
97.7	0.0026	386	Apple, juice - babyfood	1
97.6	0.0025	393	Apricot, juice- babyfood	1

Table IX. Detailed Results Dietary – Adults, 20-49 Years

Adults	20 to 49 years
Number in Subpopulation	33,538
Number of Exposure-Days	12,241,370

CARES Refined Assessment (Residues: Actual Values)				
Percentile (%)	Exposure (oral equivalents; mg/kg/day) (Weighted, per Capita)	MOE (Weighted, per Capita)	Foods resulting in exposure > 0.01 mg/kg/day	Number of Occurrences
			Strawberry	383
99.9	0.00274	365	Beet, garden, tops	142
99.8	0.00200	500	Blueberry	87
99.7	0.00160	624	Grape, wine and sherry	47
99.6	0.00136	734	Pepper, non-bell	34
99.5	0.00121	828	Strawberry, juice	22
99.4	0.00107	931	Raspberry	20
99.3	0.00101	992	Bean, cowpea, succulent	17
99.2	0.00093	1,074	Olive	12
99.1	0.00085	1,179	Nectarine	9
99.0	0.00080	1,243	Peach	9
98.9	0.00076	1,315	Cherry	3
98.8	0.00073	1,363		
98.7	0.00071	1,416		
98.6	0.00068	1,481		
98.5	0.00065	1,531		
98.4	0.00064	1,569		
98.3	0.00062	1,609		
98.2	0.00060	1,657		
98.1	0.00058	1,711		
98.0	0.00056	1,772		
97.9	0.00055	1,823		
97.8	0.00053	1,884		
97.7	0.00052	1,929		
97.6	0.00051	1,980		

Table X. Carbaryl Aggregate Risk Assessment Summary Results - Drinking Water

CHILDREN, 1-2 years

Receptors	Children 1 to 2 years		
Subpopulation	3,367 in CARES Reference Population		
Number of Exposure Days	1,228,955 (= 3,367 x 365)		
Percentile (%)		99.9^{th}	99th
Rank Order in Sorted Output		1,230	12,291
Daily Exposure (oral equivalents; mg/kg/day)		< 0.00001	< 0.00001
MOE		> 100,000	> 100,000

ADULTS, 20-49 years

Receptors	Adults 20 to 49 years		
Subpopulation	33,538 in CARES Reference Population		
Number of Exposure Days	12,241,370 (= 33,538 x 365)		
Percentile (%)		99.9^{th}	99th
Rank Order in Sorted Output		12,243	122,415
Daily Exposure (oral equivalents; mg/kg/day)		< 0.00001	< 0.00001
MOE		> 100,000	> 100,000

Table XI. Detailed Results Residential – Children, 1-2 Years

Receptors	Children 1 to 2 years		
Subpopulation	3,367 in CARES Reference Population		
Scenario Probabilities for Carbaryl	Lawn Care – Broadcast	0.209	(= 0.0042 x 50[a])
	Pet Care	0.245	(= 0.0049 x 50[a])
CARES Run Description	Probabilities were multiplied by 50 to represent 168,350 (= 3367 x 50) children[a]		
Number of Events		402	Concentrate, Hose-end Spray
	Lawn Care – Broadcast	739	Concentrate, Handwand Spray
		265	Granular, Push Spreader
	Total Lawn Events	1,406	
	Pet Care	799	Dust
Scenario Co-occurring Days		11	Two Scenarios
Number of Exposure Day	61,447,750 (= 168,350 x 365)		
Percentile (%)	99.9[th]	99th	
Rank Order in Sorted Output	61,449	614,479	
Daily Exposure (oral equivalents; mg/kg/day)	0.0001	< 0.00001	
MOE	8,398	> 100,000	

[a] The Scenario/AI Probability adjustment (50-fold increase) was only included in the "residential-only" simulations to artificially increase the number of "person-day exposure" estimates occurring in the CARES residential module output for purposes of more robust scenario-specific event and co-occurrence evaluation.

Table XII. Detailed Results Residential – Adults, 20-49 Years

Receptors	Adults 20 to 49 years		
Subpopulation	33,538 in CARES Reference Population		
Scenarios	Lawn Care - Broadcast, Vegetable Garden Care, Ornamental Plant Care, Tree Care, Pet Care, Lawn Care - Spot (custom)		
Scenario Probabilities (for Carbaryl)	Lawn Care – Spot		0.0042
	Vegetable Garden Care		0.0412
	Ornamental Plant Care		0.0484
	Tree Care		0.0112
	Pet Care		0.0049
	Lawn Care – Spot		0.0081
Number of Events	Lawn Care – Broadcast	87	Concentrate, Hose-end Spray
		172	Concentrate, Handwand Spray
		43	Granular, Push Spreader
	Total	302	
	Vegetable Garden Care	1,913	Dust
		168	RTU Spray
		77	Concentrate, Hose-end Spray
		604	Concentrate, Handwand Spray
	Total	2,762	
	Ornamental Plant Care	976	Dust
		224	RTU Spray
		74	Concentrate, Hose-end Spray
		325	Concentrate, Handwand Spray
	Total	1,599	
	Tree Care	107	Dust
		108	RTU Spray
		35	Concentrate, Hose-end Spray
		532	Concentrate, Handwand Spray
	Total	782	
	Pet Care	166	Dust
	Lawn Care – Spot	466	Dust
		62	RTU Spray
	Total	528	
Scenario Co-Occurring Days	69	Two scenarios	
Number of Exposure Days	12,241,370 (= 33538 x 365)		
Percentile (%)		99.9^{th}	99th
Rank Order in Sorted Output		12,243	122,415
Daily Exposure (oral equivalents; mg/kg/day)		0.0002	< 0.00001
MOE		4,259	> 100,000

(caneberries) that have traveled through the channels of trade and been washed before analysis. Likewise generation of a cooking factor for peaches would certainly show significant reduction in residue levels for all cooked peach commodities. Cooking factors for many commodities are not included in the assessment because they have not been specifically generated for that commodity. However numerous literature studies on many fruits and vegetables consistently show that cooking and home preparation reduces the carbaryl residue levels significantly. Finally, no attempt was made to incorporate a more realistic estimate of percent crop treated for carbaryl. Every commodity uses the maximum percent crop treated estimates from BEAD. This results in a year in which every crop has high pest pressure and is theoretically treated with carbaryl at the highest application rates, which represents an extremely unlikely scenario. Therefore, expected dietary exposure to carbaryl is probably even lower than estimated in the refined assessment described in this chapter.

Other noteworthy sources of conservatism (i.e., the overestimation of effective daily exposure levels and the underestimation of associated aggregate MOEs) and important considerations regarding interpretation of the carbaryl aggregate assessment results include the following:

1. All households using carbaryl-based products have one to two year old children (e.g., the 12-month REJV survey indicates that only a small percentage, approximately 8% (non-weighted value), of the 1.23% of U.S. households using carbaryl on the lawns, have children in their homes between the ages of zero and seven years);

2. All product users / exposed individuals re-enter treated areas every day until residues decline to non-detectable levels. The CARES residential module conservatively assumes that if a product use occurs inside or outside of a given individual's household, and is relevant to the exposure scenario, (e.g., lawn broadcast treatment), then that individual re-enters the treated area for some duration of time, on the day of application and everyday thereafter, until environmental residues have declined to essentially zero;

3. The assessment does not account for the rapid reversibility of cholinesterase inhibition produced by carbaryl (7);

4. The assessment does not account for the rapid metabolism and elimination of carbaryl (19); and

5. The contribution to the population-based aggregate exposure estimates from drinking water is overestimated since drinking water sources include not only vulnerable surface water sources as used in this assessment, but surface and ground water sources that are not vulnerable.

It is important to acknowledge that the sensitivity of other types of uncertainty analyses, while not presented in this chapter, can be conducted to determine the relative significance of selected input variables on the results of the probabilistic aggregate exposure simulations for carbaryl at selected percentiles. For example, confidence intervals for selected exposure percentiles (e.g., 99.9[th]) can be developed from the output of repeat or multiple model simulations and can provide an indication of the stability of the exposure estimates. In the case of the dietary assessment, the food form frequency analysis presented in Tables VIII and IX represent a simple type of uncertainty analysis that provides an overall indication of the key contributors to dietary exposure.

Based on the data sources used in this case study and the resulting per capita aggregate exposure and risk estimates, it can be concluded that there is "reasonable certainty of no harm" associated with the currently registered consumer product uses of carbaryl and the potential dietary (food- and water-related) aggregate exposures that may occur. This conclusion can be further investigated by consideration of the available human population-based biological monitoring data for carbaryl, as estimated from 1-naphthol levels in urine. These data reflect aggregate exposures to carbaryl and other chemicals that result in 1-naphthol excretion in the urine. Two relevant sources of biological monitoring data in the case of carbaryl include the EPA's National Human Exposure Assessment Survey (NHEXAS) data sets for children in Minnesota (see http://www.epa.gov/heds/), and the U.S. Center for Disease Control's Second National Report on Human Exposure to Environmental Chemicals, January 2003 (see http://www.cdc.gov/exposurereport/). These biological monitoring data indicate that body-weight normalized, carbaryl-equivalent dose levels are comparable to the levels predicted using CARES (dietary and residential) at and below the 99.9[th] percentile (approximately ≤ 10 µg/kg for children in the 1-12 year age range). It is also useful to consider "focused" comparisons of available "situational" or exposure scenario-specific biological monitoring data, to estimated residential exposure levels among modeled individuals, i.e., the subset of individuals who were assigned non-zero residential exposures in the CARES aggregate exposure simulation. For example, the CARES simulation results for three to five year old children indicated that the upper-percentile (99.99[th]) estimate of absorbed dose was approximately 90 µg/kg, resulting largely from lawn broadcast re-entry exposures. This upper-percentile estimate can be compared to the highest dose level (i.e., 61 µg/kg) observed among children (4-12 years) who participated in a recent carbaryl residential biomonitoriong study (see chapter by Lunchick et al.); this was the highest child dose level observed in the study's Missouri cohort where the application rates were comparable to the maximum rate used in the CARES residential lawn care simulation, i.e., approximately eight pounds of carbaryl per acre of lawn.

In summary, probabilistic exposure and risk analyses can provide a very important methodological approach for quantifying multi-source aggregate exposures within defined populations. Further, the probabilistic methods can evaluate aggregate (and cumulative) exposures to individuals within a reference population in a manner that represents demographic, geographic and temporal specificity. However, these analyses are only as robust as their underlying data sets, prompting the need for appropriate documentation, transparency, careful evaluation and disclosure of qualitative and quantitative sources of uncertainty, and whenever possible, evaluation of model estimates with relevant, direct measures of aggregate exposure or absorbed dose using validated biological monitoring methods.

References

1. Tulve, N.S., J.C. Suggs, T. McCurdy, E.A. Cohen Hubal and J. Moya. 2002. J. Expo Anal Environ Epid. 12:259-264.
2. U.S. EPA (U.S. Environmental Protection Agency). 2002. Memorandum from Felicia Fort re: Carbaryl: Revised Dietary Exposure Analysis for the HED Revised Human Health Assessment. April 28, 2002. DP Barcode D281419. Office of Prevention, Pesticides, and Toxic Substances.
3. U.S. EPA (U.S. Environmental Protection Agency). 2000. OPP Comparison of Allender, RDFgen, and MaxLIP Decomposition Procedures. Washington, DC.
4. USDA (U.S. Department of Agriculture). 2002. Conversation between Jennifer Phillips of Bayer Crop Sciences and Therese Murtagh of USDA. September 2002.
5. Nandihalli, U., S. Movassaghi, and F. Daussin. 2002. Surface Water Monitoring for Residues of Carbaryl in High Use Areas in the United States, September 11, 2002. Study No. 99S16381. Sponsor: Bayer CropScience.
6. RTI (Research Triangle Institute). 1992. National home and garden pesticide use survey final report, Executive summary, prepared for: U.S. Environmental Protection Agency, Office of Pesticides and Toxic Substances, Biological and Economic Analysis Branch, contract no. 68-WO-0032.
7. Ecobichon, D.J. 2001. Carbamate Insecticides. In: Handbook of Pesticide Toxicology – Agents. Krieger, R., ed., Academic Press, San Diego, CA.
8. Cutler, N.R., Polinsky, R.J., Sramek, J.J., Enz, A., Jhee, S.S., Mancione, L., Hourani, J., and Zolnouni, P. 1998. Dose-dependent CSF acetylcholinesterase inhibition by SDZ ENA 713 in Alzheimer's disease. Acta Neurol. Scand. 97: 244-50.
9. Jhee, S.S., Fabbri, L., Piccinno, A., Monici, P., Moran, S., Zarotsky, V., Tan, E.Y., Frackiewicz, E.J., and Shiovitz, T. 2003. First clinical evaluation of ganstigmine in patients with probable Alzheimer's disease. Clin Neuropharmacol. 26: 164-9.

of ganstigmine in patients with probable Alzheimer's disease. Clin Neuropharmacol. 26: 164-9.

10. Giacobini, E., Spiegel, R., Enz, A., Veroff, A.E., and Cutler, N.R. 2002. Inhibition of acetyl- and butyearyl-cholinesterase in the cerebrospinal fluid of patients with Alzheimer's disease by rivastigmine: correlation with cognitive benefit. Neural Transm. 109: 1053-65.

11. Unni, L.K., J. Radcliffe, G. Latham, T. Sunderland, R. Martinez, W. Potter, and R.E. Becker. 1994. Oral administration of heptylphysostigmine in healthy volunteers: a preliminary study. Methods Find Exp Clin Pharmacol. 16: 373-6.

12. Sramek, J.J., Block, G.A., Reines, S.A., Sawin, S.F., Barchowsky, A., and Cutler, N.R. 1995. A multiple-dose safety trial of eptastigmine in Alzheimer's disease, with pharmacodynamic observations of red blood cell cholinesterase. Life Sci. 56: 319-26.

13. Thal, L.J., Fuld, P.A., Masur, D.M., and Sharpless, N.S. 1983. Oral physostigmine and lecithin improve memory in Alzheimer disease. Ann Neurol. 13: 491-6.

14. Vandekar, M., Plestina, R., and Wilhelm, K. 1971 Toxicity of carbamates for mammals. Bull. World Health Org. 44, 241-249.

15. CDPR (California Department of Pesticide Regulations). 1997. Propoxur risk characterization document. Medical Toxicology and Worker Health and Safety Branches Department of Pesticide Regulation California Environmental Protection Agency, January 2, 1997.

16. U.S. EPA (U.S. Environmental Protection Agency). 1999. Revised occupational exposure and risk assessment regarding the use of oxamyl. *Barcode D263856.*

17. Krieger, R.I., South, P., Trigo, A.M. and Flores, I. 1998. Toxicity of methomyl following intravenous administration in the horse. Vet. Human Toxicol. 40, 267-269.

18. Tobia, A.J., Pontal, P-G., McCahon, P., and Carmichael, N.G. 2001. Aldicarb: Current science-based approaches in risk assessment. . In: Handbook of Pesticide Toxicology, RI Krieger ed., Academic Press, San Diego, pp 1107-1122.

19. O'Malley, M., Smith, C., O'Connell, L., Ibarra, M., Acosta, I., Margetich, S, and Krieger, R.I. 1999. Illness associated with dermal exposure to methomyl. California Environmental Protection Agency, Worker Health and Safety Branch publication #HS-1604, Sacramento, CA.

20. Ross, J.H. and J.H. Driver. 2002. Carbaryl Mammalian Metabolism and Pharmacokinetics. infoscientific.com, Inc. Document prepared for Bayer CropSciences and submitted to EPA/OPP. MRID# 45788502.

21. Somani, S.M., and Khalique, A. 1987. Pharmacokinetics and pharmacodynamics of physostigmine in the rat after intravenous administration. Drug Metab. Dispos. 15: 627-633.

22. Kosasa, T., Kuriya, Y., Matsui, K., and Yamanishi, Y. 2000. Inhibitory effect of orally administered donepezil hydrochloride (E2020). A novel treatment for Alzheimer's disease, on cholinesterase activity in rats. Eur. J. Pharmacol. 389: 173-179.

23. U.S. EPA (U.S. Environmental Protection Agency). 2003. Interim Registration Eligibility Decision for Carbaryl (List A, Case 0080). June 30, 2003. U.S. Environmental Protection Agency, Office of Pesticide Programs, Washington, D.C.

24. U.S. EPA (U.S. Environmental Protection Agency). 2003. Carbaryl: Revised HED Risk Assessment – Phase 5 – Public Comment Period, Error Correction Comments Incorporated; DP Barcode D287532, PC Code: 056801. March 14, 2003. (J.L. Dawson).

25. Ross JH, J.H. Driver, R.C. Cochran, T. Thongsinthusak, and R.A. Krieger. 2001. Could Pesticide Toxicology Studies Be More Relevant to Occupational Risk Assessment? Ann. Occup. Hygiene, 45 (Suppl 1): 5-17.

26. U.S. EPA. (U.S. Environmental Protection Agency). 1998. Memorandum from Lori Brunsman to Virginia Dobozy re: Revised Carbaryl Quantitative Risk Assessment (Q*) Based on CD-1 Mouse Dietary Study with ¾'s Interspecies Scaling Factor. October 28, 1998. Office of Prevention, Pesticides, and Toxic Substances.

27. U.S. EPA. (U.S. Environmental Protection Agency). 2001. Organophosphate Pesticides: Preliminary OP Cumulative Risk Assessment – Residential Chapter. http://www.epa.gov/pesticides/cumulative/pra-op/ and related documentation http://www.epa.gov/pesticides/cumulative/pra-op/hedinput.htm.

28. Wilkinson, C.F., G.R. Christoph, E. Julien, J. M. Kelley, J. Kronenberg, J. McCarthy, R. Reiss. 2000. Assessing the risks of exposure to multiple chemicals with a common mechanism of toxicity: how to cumulate? Reg Toxicol Pharmacol 31:30-43.

Indexes

Author Index

Subject Index